MYSTICAL ANTHROPOLOGY CONSTITUTED AS A FUNDAMENTAL VITALOGICAL SCIENCE

by Dr. Johnny Lovewisdom

INTERNATIONAL UNIVERSITY OF THE NATURAL VITALOGICAL SCIENCES is the educational vehicle of the earth's most Ancient Mystical Order, and Strictest Scientific Discipline in Laws for Natural Living in the Primordial Divine Design, formerly known as "The August Order of Living Immortals in Conscientious Tattwas", now legally registered as THE PRISTINE ORDER OF PARADISIAN PERFECTION of The Heavenly Ecclesia of the Living God. Dr. Johnny Lovewisdom, Patriarch-President & Dr. Ruth Marie Lovewisdom, Secretary "Build Paradises and eat the Fruits thereof!"

Disclaimer notice: This book is intended as a reference volume only, not as a medical manual. This book is sold for information purposes only and is not intended as medical advice. The information in this book does not diagnose, prescribe, treat or cure any diseases, rather it promotes general health and well-being through better nutrition, exercise and lifestyle adjustments. The information given here is designed to help you make informed decisions about your health. It is not intended as a substitute for any treatment that may have been prescribed by your doctor. If you suspect that you have a medical problem, we urge you to seek competent medical help. The dietary and exercise programs in this book are not intended as a substitute for any dietary regimen or exercise routine that may have been prescribed by your doctor. As with all diet and exercise programs you should get your doctor's approval before beginning. All forms of diets and exercises pose some inherent risk, therefore the editors and publisher advise readers to take full responsibility for their safety and know their limits. Due to the fact that there is always some risk involved, the author, publisher and distributors of this book are not responsible for any adverse effects or consequences resulting from the use of any suggestions or procedures described herein. Neither the author, the publisher, nor the distributors will be held accountable for the use or misuse of the information contained in this book.

Copyright © 2010 All Rights Reserved
Reproduction or translation of any part of this work by any means, electronic or mechanical, including photocopying, beyond that permitted by the Copyright Law, without the permission of the publisher, is unlawful.

Printed in the United States of America

Other Titles published by Paradisian Publications
and authored by Dr. Johnny Lovewisdom:

THE BUDDHIST ESSENE GOSPEL OF JESUS, VOLUME I
THE BUDDHIST ESSENE GOSPEL OF JESUS, VOLUME II
THE BUDDHIST ESSENE GOSPEL OF JESUS, VOLUME III
THE LOVEWISDOM MESSAGE ON PARADISE BUILDING
THE HEALING GOD SPELL OF SAINT JOHN

TABLE OF CONTENTS

CHAPTER I: INTRODUCTION TO HYPERBOREAN CULTURE AND THEIR SUN GOD. Seven Eternities ago the Sweat-Born came forth from the First Heaven to inhabit the earth with huge vaporous bodies in the earliest Hyperborean Race nourishing from the air, which in time descended into the frugivorous, required a living water nourishment from juicy fruits. Modern misconceptions as to seeds and nuts which give densest body, drugging with pesticides for Fruit-eater's "High", early historians identification of Hyperboreans and original source of Sun Gods.

CHAPTER II: THE ANCIENT EASTERN LEGENDS OF PARADISIAN ORIGINS. Buddhist Concepts of how self-luminous bodies became denser partaking of earth-born fruits of earth, till finally eating rice, men became passionate and evil; Comparison with Bible Genesis when man was cast out of Eden for bread-eating; Pre-Adamite Man that lived in Altai or "Heavenly Mts.", Shambhala, Paradise of Chinese Legends, P'eng lai, Taoist Hygiene School abstains from 5 grains, eats jujubes fruit, or breath, to become Immortals, Lemurians and 7 Root Races, Soma or Juicy Fruit in Hindu and Zend legend, Science on recent origin of grains, and Map.

CHAPTER III: THE HEBRAIC ORIGIN OF THE OLD TESTAMENT BIBLE. Shamash, the Sun God of Babylonian Legend, and part played in Biblic Genesis, the Legend of Moses' birth and Great Flood precisely described in Chaldean Records, Josephus explains Genesis, Adam and Eve, Cain, Seth from non-biblic sources, Gnostics were Essenes, Ezra writes Mosaic Books.

CHAPTER IV: DIETETIC VIOLATIONS THAT GAVE RACIAL TRAITS TO PREHISTORIC PAMIRS, TARIM BASIN, SUMERIA, CHALDEA FROM ATLANTIS. Anthropology and Ethnology, Iridology gives color index to pathology in the eyes and its relation to skin and hair; how skin color is developed by ethnic traits in diet; negroes become white on raw food; clabber as living substance responsible for white race; intestinal purity and alkalinity gives clean skin color; Pamirs described as original Eden; Chinese corroboration about Eden; Sumeria, fish-born Semiramis, and Atlantis.

CHAPTER V: THE EGYPTIAN INITIATION INTO ATLANTEAN MYSTERIES, THEIR SACRED HIERATIC LANGUAGE AND TRAITS. Historic Greek descriptions of Atlantis, Early Egyptians abstain from grains and flesh, Moses copies Pork and unclean meat doctrine from Egypt, Excess fat, avocados, etc. hard on liver, Lactobacillus, healing of menstruation and seminal losses, the Mysteries of Egyptian Initiation, Hermes, Atlantian writings.

CHAPTER VI: ANCIENT LEGENDARY HISTORY OF THE KRISHNA CULT, AND ARYAN CULTURAL BENEFICIENCE FROM BUDDHA. What Scriptures confirm as to the black man Krishna who creates rivers of blood slaughtering enemies by the hundred thousands, made over 4 women pregnant every night of his life, having 16,108 wives, including female gorilla, and other jungle legends unwittingly espoused by Yoga much like O.T. Bible's gory prophets. Buddha's Doctrine of Compassion for all beings, abstinence from killing and chaste ideals, forsaking worldly desires, Teachings.

CHAPTER VIII: THE FIRST PEOPLE, THE LEMURIANS, THE GOBI CIVILIZATION, ANTHROPOGENESIS OF MODERN SCIENCE AND MYSTICAL LEGENDS ABOUT SHAMBHALA. After Eternal or First Land of Hyperborea, Lemuria, sin and taking of life came into being. Seed-eating gave rise to sexualism, and perversions giving bestial forms to human fetus development. Soviet science as to Lemuria, Osborn and Doreal Research about Gobi Civilization, Tibetan Lama's views and Prophecy of Shambhala in Ecuador.

CHAPTER IX: ALL MATTER ONCE LIVED, HYPERBOREAN TRAITS, MAN'S EARTH ARRIVAL, DESCRIPTION OF ALTAI REGION, AND SACRED SANCTUARIES. Esoteric!

CHAPTER X: CONCLUSION. Genetic Custody of Hyperborean Traits and Cultural Wisdom inherent with Maitreya's Mission, restoring the Rose Apple Grove of Gods in Southern Shambhala of the High Andes, with detailed compliance to Ancient Prophecies and Paradisian Ideals of Perfection.

MYSTICAL ANTHROPLOGY
CONSTITUTED AS A FUNDAMENTAL
VITALOGICAL SCIENCE
BY DR. JOHNNY LOVEWISDOM

STARTING WITH THE ANCIENT HYPERBOREANS, THRU MILLIONS OF YEARS, WE SURVEY THE LIFE AND IDEALS OF PRE-HISTORIC CULTURES, LEADING TO OUR PREPARATION FOR A MYSTICAL RECURRENCE OF THE SATYA YUGA, OR THE NEW GOLDEN AGE, FOSTERED BY MASTER PARADISIANS OF A HIGH HEAVENLY HIERARCHY. (Preamble to our Teachings follows,)

THE SONG OF THE SOUL VICTORIOUS! (Sung by a Hyperborean Yogi many Thousands of Years Ago.)

I stand in the Great Forever,
I live in the Ocean of Truth.
I bask in the Golden Sunshine,
Of Eternal Love and Youth.
And God is within and around me,
All Good is forever mine.
To all who ask it is given,
And it comes by Law Divine.
In the deathless glory of Spirit,
That knows no destruction or fall
From the Immortal Fires of Heaven,
To the plains of earth, I call
Who is this "I" that is speaking,
Being so wondrous in might?
Tis a part of the Primitive Essence,
A spark of the Infinite Light.
Blasphemous and vain they may call me,
What matters it all to me?
Side by side we are marching onward,
In time we shall all agree.
Oh, I stand in the Great Forever,
All things to me are Divine.
I partake of the Heavenly Nectar,
Of Paradisian Fruits, I dine,
In the gleam of the shining rainbow,

The Father of Love I behold,
As I gaze on its radiant blending,
Of Crimson Blue and Gold.

In all the bright birds that are singing.
In all the fair flowers that bloom,
Whose welcome aroma is bringing
Their blessings of sweet perfume,
In the glorious tint of the morning,
In the gorgeous sheen of the night,
My Soul is lost in Rapture,
My senses are lost in Sight.
Come back, oh my soul, in thy straying,
Let thy wandering pinions be furled,
Speed back thru the Heavenly Ether,
To this posy and sense bound world.
They say I am only mortal,
Like others I am born to die,
In the mighty will of the Spirit
I answer, death I defy.
And I feel a power uprising,
Like the Power of an Embryo God,
With a glorious wall it surrounds me,
And lifts me up from the sod.
I am born to die, Oh never,
this Spirit is all of me,
I stand in the Great Forever,
Oh God, I am One with Thee.
Ye, Pilgrims of various probations,
Ye, Teacher and Saviors of men,
To our heaven-born Revelations,
My Spirit shall answer, Amen.
I think of my birthright Immortal,
And my Being expands like a Rose,
An odorous cloud of fragrance,
Around and above me flows.

A glorious song of rejoicing.
In my innermost Spirit I hear,
And it sounds like Heavenly Vices,
In a chorus Divine and clear.

Oh, the Glory and Joy of Living!
Oh, the Inspiration I feel!
Like a halo of Love it surrounds me
With newborn Rapture and Zeal.
I gaze thru the dawn of the morning,
I dream 'neath the stars of the night,
I bow my head to the blessings,
Of this wondrous Gift of Light.
Oh God, I am One forever,
With Thee, by the glory of birth,
The Celestial Powers proclaim it,
To the uttermost bounds of the earth.
Oh, the Glory and Joy of Living!
To know we are One with God:
'Tis Thy Omnipotent Power of Spirit,
'Tis a blossom that crowns the sod.
With you in the Great Forever,
By the children of earth I stand.
And this Light flowing out like a river,
Shall bless and redeem the land.
Thus I stand in the Great Forever,
With Thee as Eternities roll;
Thy Spirit forsakes me never,
Thy Love is the home of my Soul

THE PRISTINE ORDER OF PARADISIAN PERFECTION
The Most Sublime and Strictest Discipline on Earth
ALSO KNOWN AS THE AUGUST ORDER
OF LIVING IMMORTALS

To realize the strict abstinence from the Active Causative Substances that give origin to Cravings, Lust and Disease, as the Basic Source of Thoughts, Attachment and a Life in Sin:- WE FORSAKE the consuming of eggs, nuts, grains, legumes and other SEEDS, - which augment Reproductive Substance-Activities; beside fish, fowl or all animal FLESH, cooked foods, leather, pesticides, drugs, etc., - which augment Killing or destroy Sentient Life. The ideal of this Spiritualizing Dietetics, or VITARIANISM, is eating only Juicy Fruit, - supplemented with succulent vegetables, when necessary. When one achieves true Sublimation of Libido thus, thru our precise Dietetic Science of Continence, or VITALOGY, one eliminates and heals menstruation and seminal losses as well as most diseases. All Sin and Unhappiness is the Lack of Love for our Creator and for one another. When we live by loving, also as male and female, one in the Image of God, sustained by Living Water in juicy fruits of Paradise, - the ideal we live for,- we shall be Perpetually Born again from our embryonic seed if we retain it within and nourish this Agape of Love. Thus, we Forsake the World of Sin, Crime, Deception and its evil Karma to BUILD PARADISES to be with God and to Nourish from the JUICY FRUITS, HIS HEALING, SAVING AND LIVING WATER OF LIFE EVERLASTING! // ONLY THRU THE PRACTICE OF THIS PARADISIAN LIFE OF IMMACULATE CONCEPTION, WITHOUT BLOODSHED, POPULATION EXCESSES, WAR AND CONSEQUENT ECOLOGICAL PLAGUES, - CAN THE NEW AGE AND RACE SURVIVE.

FOREWORD

This, too, shall pass away, nothing in this world is infallible, and all created things return to whence they came. Both Sacred Scriptures and words spoken by Illumined Immortal Beings inhabiting mortal flesh, all manifest by passing mortal means. The meaning of words change with time nor can Truth be captivated by such a changing media.

This is why the Spiritual Guides of Mankind, too, can only speak forth relative Truth for their time and world condition, designed for their epoch in the ascending path of man's spiritual evolution. We are not concerned with promoting or degrading any special religious culture, altho it will be seen that extreme tendencies provoke contrary ends.

The Love-Wisdom Message of Maitreya, also known as the God-Endowed Healer of Mankind, in the Mystical Truth Body of the Avatar-Buddha-Christ, or possibly worded different in the reader's ideals than these expressions imply, recognizes the fact that humanity is today at the Cross Roads, having out-grown previous paths leading toward the Godhead. In direct illustration, may I state that in this New Age upon us, the long-honored "Holy Life" style practiced by Christians, Buddhists, Yogis etc. contradict virtue, purpose, if not the God-Given vocation of man.

In such monastic ways of hundreds and thousands of years past, men and women were taught to forsake the life they were living, earning the necessities of life by physical labor, usually, giving up all they had owned, to become mendicants, or religious beggars, an added burden on the public, most of which were enslaved or taxed heavily already by their wealthy rulers. These holy men could peacefully while away their time contemplating abstract ideals as to God and Sainthood, practice religious exercises in prayer, ritual, indoctrination, making idols or scriptures, etc. In the process, Religion became a weapon to frighten people into subjection. In Europe, India, Tibet, etc. as much as one half of all the population depended on being fed, clothed and sheltered by the other half of the people who did added physical labor to sustain themselves and non-working religious. Conditions have changed little in that today Science has become man's Religion, scientific research spends immense funds to sustain their faith, the earth is polluted with millions of idols, oracles, and fast growing uselessness of these devices along with economic welfare sustaining large populations of jobless and unemployables. However, unlike modern scientific times, the religions belonging to Holy Orders of Priests, Monks or Nuns, did not engender crime to the extent it is prevalent today, crime being often unknown since the multitudes sought very little in life for which they might covet life or things. Such "Dark Ages", as Science labels them, gave birth to a "Spiritual Renaissance" of cultural philosophy.

However, when in vogue, these seemingly harmless holy men were living like parasites on the backs of the working population. They did inspire spiritual ideals, we must admit, but the basis of their lifestyle was eroding or rotting away by false methods to reach desired ends. These monastic, lamasery, mendicant, etc. people were sustained almost exclusively on grains, whose cultivation eroded the fertile soil, turning their lands into sterile fields and desert. There are hundreds of other damaging practices perpetuated by "God-Inspired" men, but this well portrays the ungodly evidence in the land, beside excessively corpulent bodies of people who generally live in such grain eating communities.

This is why the Love-Wisdom Message of Maitreya, The Pristine Order of Paradisian Perfection,- as a Mystical Order exemplifying the teaching, in our professed Spiritual Guidance of a Cosmic or Higher Heavenly Hierarchy in the New Age has completely removed the ideals to be practiced in our Holy Life style from past Truths, or Principles given by Christ, Buddha and other Great Teachers. Instead of seeking an ideal of a Supernatural Life in an Artificial Way, we seek Nature's God-Given Laws for the Return to the Golden Age of Paradisian Perfection.

In this thesis, we seek to trace man's earthy origin thru millions of years, led astray by loss of purpose. The Spiritual Guides can only guide men into compliance with Universal and Natural Laws, without the threats and punishments fostered by religious and political governments. "All knowledge and arts are to be found in Nature", Hippocrates wisely said, "If we question her properly she will reveal to us the truths that pertain to each of these and to ourselves. What is Nature in operation but the very Divinity itself manifesting the presence. We must proceed with faith with firm assurance of discovering at last the whole truth, and nature will let us know her answer, thru our Inner Sense"......

CHAPTER I: INTRODUCTION TO HYPERBOREAN CULTURE, AND THEIR SUN GOD

Beloved Brother Sister in Divinity:

Recall our mutual state In Immortality! Remember the Sunny Days of our childhood aeons and aeons ago. We all still have, are and will be just One Universally as we were Seven Eternities ago, even after Boundless Absence of Light, Creation and Existence. Yet, let me not weary thee with Transcendence, and speak as I must of our recent status as visible beings even if this tends to flirt with the tangible limits of Spiritual Liberation by creating form.

In this world governed by the Seven Creative Spirits, or the Great (Maha) and Dhyani Chohans, as already described in the most ancient Esoteric Manuscripts, we can only speak within the limit of beings, altho the "Corporal cannot express the Incorporeal", as Hermes told us, or that "Reality is not upon the earth". What is subject to being created or birth, is not Real, for it exists by change, conditioned by time. So briefly, we return from the First Heaven, and our Resplendent Father-Mother Immortals also of Fleshless Being or Spirit and Truth of Reality, to this world of Will Born Lords, the Chohans, to the Second Race, a product of expansion, even the asexual from the Sexless Self-Born. And even those Sweat-Born Beings would bear a Third, the Egg Born.

In the "Secret Book of Dzyan" in its ANTHROPOGENESIS, Chapter VII, 27 in Helen P. Blavatsky's "SECRET DOCTRINE" text, it tells us: "The Third Race became the Vahan of the Lords of Wisdom. It created `Sons of Will and Yoga', by Kriyasakti it created them, the Holy Fathers, Ancestors of the Arhats." The "Lords of Wisdom" are the Buddhas (Illumined or Enlightened Ones) existing from the beginning of the world. Now, the Ancients held that any idea will manifest itself externally, if one's attention and will is deeply concentrated upon it. This is the holy seed of a Higher Mankind, the future Saviors and Leaders of the New-Age Race of PARADISIANS. One may well understand the intuitional work I had to perform as the Maha Chohan or Spiritual Father of the New-Age Race, and how the human race has fallen into disgrace and degeneration because carnal men and women believe they can better the race by lower mammal means of reproduction.

This was told of by most ancient Scriptures, as to the Fourth Race, who were the children of Padmapani (Chenrezi). Chenrezi (Padmapani or Avalokiteswara are its descriptive names) is the Tibetan Logos that the Tashi and Dalai Lamas were supposed to incarnate as the "most perfect Buddha". The long prophesied incapacity, manifest with the broken lineage at the time of the 13th Lama, made for the Esoteric Mission of the True Lord of Wisdom, Dhyani Cherenzi, to manifest physically by the thus named one coming to the High Andes to initiate the Maha Chohan of the New Age Race, and consequent formulating of the PRISTINE ORDER OF THE PARADISIAN PERFECTION as the HIGH HEAVENLY HIERARCHY and our work creating the Paradisian Race by Divine Will and Yoga, or Divine Oneness in Spirit and Truth, rather than by carnal reproduction. The non-mammal reproduction process is achieved by uniting of drops of perspiration, which gather into greater and greater drops until they become ovoid bodies or a huge egg which with warmth of solar incubation developed gigantic seemingly vaporous bodies. Thus were the progeny of Androgynous man, of both male and female equilibrium, who were the first Hyperboreans who lived without eating, having a huge vaporous body like clouds. But partaking of juicy fruits of the trees that clouds water, beings with definite human form were created by the Will or desire to become something tangible.

   This then is the story of the Sage Kandu, who was tempted to spend 907 years, 7 months and 3 days in Ecstatic Trance with a beautiful girl nymph. The nymph went from tree to tree drying her limbs from the living dew of her perspiration, which the cool breeze like a wind united by breathing in and blowing out the breath of Life, and the rays of Soma (the Moon) matured and grew into an egg-shaped fetus. The newborn grew to reach the height of tree tops, to become the lovely girl, Marisha.

   Marisha, like the mother of Krishna, lived on fruits alone before the realization of Immaculate Conception, which was told of allegorically in the birth of Siddhartha Gautama who came to be the Buddha, and John the Illumined (Baptist) who became the Savior or "Jesus", the Christ.

   Kandu symbolized the First Race, devoid of Mind or an

Ecstatic Trance for nearly a millennium. Mind came only after eating of the juicy fruits of the Garden of Paradise, just as Reproduction came by feeding from reproductive substance, or plant seeds. The true fruits or juicy pulp invite man to spread the seeds of plants that mankind desires for his Paradise on earth. Juicy apples are really the brain food of the Gods and Goddesses of Wisdom, and the golden, peaches, apricots and nectarines are food of the Immortals, and not the almonds man developed from cultivating the seed of stone fruits till they no longer had any juicy pulp around the seed. Walnuts are 2.5% water, Almond butter is 2.5% water and Pecans 3.4. In turn the juicy pulp of Apples is 86% living water, Peaches are 88% water and even a tomato is 94% water. In contrast the HIGH PROTEIN content of Reproductive Seed Substances, also match their dryness, Black Walnuts being 27% Protein, Soybeans 34%, Sesame Seed 36%, Wheat 14%, Almonds 21%, while lack of Protein necessary for dense heavy humans and beasts is observed in the low protein content of Apples, 0.4%, Apricots, 1.4%; Peaches 0.7%, Nectarines 0.6%, etc.

    Now, with a Mind and Juicy Fruits, man need not work too hard when he has provided himself with an Orchard or Grove of Fruits, enabling the development of Wisdom, freeing one from ignorance, which begets child-birth, life and death, the lot of earthlings in a materialistic world. Fruits are directly assimilable, doing away with the need of many difficult digestive processes, and taxing the liver and kidneys which give root to all kinds of painful states of diseased flesh.

    But, Alas, the very keys to the Royal road to Recovery of Paradise Lost, at Last, requires the same process as in the First: The Paradisian Enlightened INITIATION of the Baptism, Purification of the Living Water of Juicy Fruits renewing every cell, along with the Solar and Lunar Purification of the Holy Spirit of Heavenly Father and Mother Nature in the work of Building Paradises to earn those fruits worthily. Only years of Sweat-Born Labor, like the Virgin Birth of Marisha watering fruit trees with Sweat, Tears and Painful Labor are required to have any fruit diet that one can ever hope to Regenerate one's form to gather together the Ovoid Karmic Egg of True Paradisian Virtues. To work we need strong bodies able to build the required Paradise Regained, and a strong mind that will not be tempted into suicidal means to arrive at a purely Spiritual Status, so there is a need of concentrations of the organic minerals, enzyme proteins, vitamins beside fruit carbohydrates.

To work, we must have a strong skeleton, solid flesh, muscles and lungs to oxidize food wastes and take in solar energy, all of which in consequence takes a heavy material body. To digest food we need digestive organs, stomach, intestines, liver, etc. To get this food requires the body frame work, muscle motor power, and a brain computer and nerve sensor mechanisms, developed by Life in this Dense Material World. But remember also that the First Race became the INTERIOR (or Etheric Double) of the Second or Hyperborean Race of Gigantic Rarified substance and form abiding in the prehistoric Arctic Continent of Tropical warmth.

The High Heavenly Hierarchy, then as now, understood why, just how one living on juicy fruits without assimilating reproductive seed substance eliminates menses and seminal losses and consequent contagion of suffering in other flesh-born victims of desire. Love and Ecstasy need none of the mammal heat of passion, raping a mate when one's mentality cannot overcome passionate compulsion, but becomes enlightened Kindness. Population excesses, food lack and excesses, labor disputes or exploitation, war and politics, etc. all become superfluous in the Garden of Eden. Good and evil character are no longer the problem with the ideal type in Homo Paradisiacus, so sexual reproduction giving variables in race and character are no longer needed to give enslaved members of man and woman kind, beasts of burden, and lower animals that do away with plagues that unbalance the earth's ecology, and all of Sexual Creation.

Trees and plants needed to exist to feed humans and animals, and man and animal organisms are needed to keep and maintain the Garden of Paradise,- which has now degenerated into thorns and fruitless trees that bear seeds alone, while animals and men destroy each other in a Savage Struggle of the Fittest in bestial passion and suffering, all because former Paradisians were tempted to eat of the Tree of Knowledge of Good and Evil,- Reproductive food substance that gives Reproductive Proclivity. Once Evil came into the world, it increasingly contaminated everything with its Pathological Contagion, even the Good with Suffering.

As long as we have a strong solid body and an intelligent brain, humans need to eat the juicy fruits and plants of the earth. Without any solid body or human form, floating as a vaporous cloud, we can dispense with earthy fruits and plants. Clouds water trees, that humans may eat the fruits thereof to consume the Living Water, in which we resemble a heavenly nymph who dried herself (of sweat and tears) on the Paradisian trees, which in turn beget Paradisian Yogis like Kandu who live in Ecstatic Contemplation of Samadhi or Union with Divine Design and Will. Our brain is 81% Living Water, our blood 90%, our muscles 75%, therefore we contemplate really a vaporous celestial being that human bone marrow is distilling into blood cells and lymph that fashions human structures. The Paradisian Immortal needs not replace a continuous form ever ascending higher, with flesh born offspring, Omniscient in Higher Law/Life.

Now, people say, however, if the whole earth became perfect, there would be no one to reproduce the race, without menses and seminal losses among humans. In that case, there would be no one to damage our environment, death would become obsolete, so we would need no one to replenish even the present population excesses, even if two-thirds of mankind died in the process of adaptation that might arise due to ignorance in methodology. Paradisian giants estimated to have been 22 feet tall, needed no stepladders to pick fruit from trees, nor muscular bodies to do heavy work, and having even a thousand years of Life, in turn would require 100 years of adolescence, giving time for studying vast libraries of information, enabling them to develop living computers of brain-power so that it would eliminate the school of painful experience that dense humans inherit in a short-lived world of weak-willed flesh.

Modern science is already clumsily invading facts of Hyperborean Wisdom when it speaks of androgynous reproduction without the mammal type of mating. By cloning, scientists seek to isolate the reproductive cellular genes, since our bodies have both male and female components in order to restore damaged cells with new ones, and thus use them to develop Life "in a test tube", just as already they have made "test tube babies". Such androgynous, virgin birth or parthenogenetic methods remove chance disappointments, since it is like budding or grafting branches of desired quality fruit trees which continue to produce the desired quality of fruit. This Virgin Birth method would produce identical children from their mother, just as in theology, a Spiritual Virgin Mary, "Mother of God", was needed to be born again at the Bethany Crossing of the Jordan to give the Son of God, not born of flesh. Love with mating of mortal flesh never made perfect Sons of God. Idealist fruitarians, etc. can only produce children who carnally inherit their own inability to face the ever-increasing adverse realities of life on earth now, and thus propagate their own helpless dilemma in their children, who in turn have a thousand-fold increasing challenge in obstacles from the population explosion of non-idealist, brutarian, etc. or violent revolutionists needed in ecological birth checking. As Wise Buddhas have taught, birth begets birth, with suffering and death too.

Biogenic Stimulators are found wherever there is struggle for life and adaptation, V. P. Filatove of U.S.S.R. observed (Tissue Therapy). Therapeutic effects are derived from mud-baths, storing green leaves in dark places, sauerkraut, clabber and other milk cultured acidophilus bacterial plants, cutting of tree bark or grafting to give fruit early, hard physical labor, high altitude, etc. which stimulate survival factors. However, when this factor is abused to an excess, there comes an end point giving intoxication and fatal destruction. Just as seed cooling prior to planting promotes frost resistance, observes I. I. Tumanov, A. Astaurov showed that a short period of sub-lethal high temperature results in artificial parthenogenesis in experiments with eggs of silk worms. This shows the relationship in why sun baths are regenerative (unless overdone giving sunstroke and cancer). Finnish people owe many early Olympic Championships and racial health to the hot Sauna Baths. To sum this up, we drive the Virgin Birth Legend of the Sun God Apollo or Helios, the prototype among later legends decorating the lives of Krishna, Buddha, Christ, etc. Even the Pre-Inca Hyperborean legend of the androgynous Sun God, and Son of Inti, gives a name of all Indians. Enzymes are the factor in genes that transmit heredity. The fall of temperature makes enzymes inactive. Taking ice cream or frozen dessert, or cold water packs on the digestive areas can force one to have bowel movements 3 or more times a day, and equalize body temperatures as some "therapeutists" recommend. With a rise in temperature, the enzymes are activated resulting in metabolism, while excessive fever in digestive organs, resulting from hot cooked foods or incompatible combinations, in turn can result in cooking whatever enzymes are left in food (if any), beside resulting in fermentation and intoxication of the organism. In the same manner, if bestial mating of mammals became obsolete, both the male and female genes and enzymes of human generation could be awakened in Paradisian Sun Gods/Goddesses enabling the regeneration of the New Age Race, just as the Parthenogenesis took care of the Hyperborean demographic ecology, and their single eye in the forehead could have given them an insight to future suffering, if heeded, with Omniscience. The sweat-born of those mating without mammal, bestial principles of Regeneration, like even the egg-born parthenogenesis could become true reality when humans recover Hyperborean secrets of those "Shaman" Alchemists ages or aeons ago in once tropical Siberia.

The same would apply to the secret transmutation of one element into another by the Living Water Content enabling everlasting life, by eliminating the kind of mineral, vitamin and enzyme concentrations that seeds (cooked flesh, etc.) furnish. Instead, humans will look to the leaf protein enzymes that herbivorous eat to have powerful bodies able to rebuild earth into one continuous Paradise, and the rarified factors that in fruits perpetuate Omniscient beings. In future chapters of this Mystical Anthropological Science we shall describe man's prehistory.

Those who read Volume I of our "MAITREYA" book, have already had a glimpse of how seriously I went into trying to maintain a body on strictly juicy pulp of fruits or Living Water, followed soon with vain attempts of giving up even juicy fruits living on 99% pure water. Of course, these ideals have an aesthetic quality of beauty, until one of solid body and bones realizes that I had tried to leap from earth to heaven, without developing a vast Paradisian Plane of Being and Experience as the Father, or

Exemplar to followers, that would require building Pure Paradises before talking about an exclusive diet from the juicy fruits thereof, let alone a bodiless and foodless Life-Plane.

From the bodies and environment of earthy worldlings we do not automatically inherit instant ready-made Paradises. This is only the inheritance of complete Spiritual Rebirth, a Mystical and Theopathic realization, so that one does not remain an earthbound worldling, flesh bound to a body that is genetically passionate, desiring and unconsciously ever seeking psychosomatic returns to the old pattern and civilization's protective comforts. Biogenesis requires natural re-adaptation.

Today's Paradisian is in great danger of early death trying to live on juicy fruits without his required Initiation of having Built True Fruitarian Paradises to recognize when one errs. Thousands of fruit-eaters, whether inspired by earlier experimental failures following erring leaders, or even the advance theories of my earlier experience told of in Spiritualizing Dietetics/Vitarianism book, suddenly decided they too were Paradisians ready to abandon civilization, to run everywhere looking for ready-built Paradises to live in, or sooner or later ending up living off of fruits from super-markets or commercial growers since only the wealthiest and least greedy for money could afford the so-called organic foods. Death and degeneration became synonymous with experiments with strict fruitarianism. Moreover, none, not one ever achieved even a life drinking only 99% pure water, let alone adapting successfully to juicy fruits, because they did not have the corresponding SPIRITUAL REBIRTH acquired in a Contemplative life seven years as a hermit in the High Andes without even a fruit diet, civilized protective comforts, and "penances" that develops biogenic re-adaptation.

First of all, even with a Spiritual Rebirth, without a paradise of pure fruits in a dry, warm California-like conducive climate (for which I had abandoned cold, humid Western Washington, Paradisian home with lots of juicy fruits), I became dependent on heavily poisoned fruits and vegetables of city markets or agro-factory farms artificially grown. This is the lot of most all would-be

Paradisian Fruitarians. Hybrid fruits cultivated for dessert dishes of the Omnivorous wealthy are not body building, due to, first, the unnatural cravings of civilized appetites seeking to balance added condiments, sugar, meat, grain, drugs, etc. To speak candidly, Fruitarians, natural Hygienists, etc. are mostly NOTHING BUT DRUG ADDICTS, getting "High" from Chlordane, Parathion, Toxiphene, and a hundred other deadly poisonous drugs used by the Agro-Business Factories to cure plants so nothing living can survive on them including insects, fungus and human beings. Every item has some in it. Just as I was, fruitarians are reluctant to discover how the fruit they eat is grown. As vegetarians one talked of the need that vegetarianism opponents had for visiting slaughterhouses, hog-raising lots, etc. to see vegetarian opinions and beliefs. How many fruitarians who are not at all reluctant to tell even vegetarians about the garbage, slop, etc. they are eating, feeling so superior in eating just juicy fruit, in turn, have witnessed first-hand working in fruit farms, or lived with an agro-business family to learn how periodically the fruit trees are drenched with the most potent drug poisons, sprayed on the trees, absorbed by the roots, contaminate the water, kill the weeds, fertilize the trees, and observed cancer, paralysis, neurosis, and other victims now living in hospital care or dead already? Even when one has observed such things, like tobacco, alcohol, etc. user,- they prefer to cover their own case up, believing they will come out as exceptions!

These Agro-Chemicals not only destroy the nerves, cause tumors, attack both liver and kidneys, create mental depression, indigestion, beside remove all the friendly acidophilus bacteria in the intestines so fruitarians assimilate very little, if any, of the fruits they do eat, ever with food-cravings for something more substantial. Yet one remains like Kandu in his hypnotic trance, unable to see the horrible crimes they commit each time they consume death-laden foods, destroying the cellular tissue of their body. City dwellers find they must remain within their sterile fumigated and polluted atmosphere to enjoy a life sunbathing, not able to tolerate gnats, mosquitoes, flies, etc. especially attracted to drug users in pure country air. There are a hundred and one ways that people's lives evade Paradisian Principles, other than diet, that can seriously damage health and well-being. Yet, there is no easy road, but it requires years of re-adaptation to a natural primitive life, distant or isolated from civilized people, supermarkets, health food pharmacies, etc., to grow one's own fruitarian Paradise. It still takes hard farm work beside acquire the corresponding "being-alone" meditational mental and Spiritual Development witnessing the need for the body to become integrated with the Cosmic Being we are within a pulsating cell of our environment.

Well do I remember how a master in our circles, Bill Goodell, said he felt in the holy Presence of a 750 year old Jain Master-Teacher, whose gaze pierced one's every thought and opened up one's life history leaving one with no way to cover up, which most conversation among the so-called "sociable" people really amounts to. When one does achieve any status among PARADISIAN MASTERS, very few can stand their Darshan, or Presence. Either the disciplined life without cover-ups and alibis so disorganizes the visitor's thinking processes, so he forgets the many inquiries he had arrived with, or a neophyte leaves complaining that such Masters are too hard to get along with, since one's position is too humble to bear, after so long having assumed oneself as the only righteous one among one's associates, and there is no common sociable attitude of ignorance or neglect that both parties can seek refuge under. Paradises do not come by crying and kicking, in the suffering one endures in one's undisciplined life. Eating fruit or fasting solves nothing. "Searchers" are unbelievers, only little better off than skeptics.

The foregoing plea for prudence in one's proceedings is given to caution those who feel they must eat only juicy fruit, write a book that compromises our principles with their ideals and failings, or some other cover-up to rest their minds. Life enjoying Paradisian Bliss only comes by making one's surroundings Paradisian or BUILDING PARADISES.

To start our detailed Prehistory of the HYPERBOREANS, may I warn the student that there is no special reason why we should call our ancient origins Hyperborea, since that is Greek and not of their original tongue. We have chosen Hyperborean simply because in the West we refer to Herodotus as the "Father of History", and he described us as Hyperboreans. Herodotus was reluctant to report on the Hyperboreans who bordered on the Arctic, remarking "Neither the Scythians nor any of the neighboring people, the ISSEDONES alone excepted, have any knowledge of those Siberian Ugrian tribes, and what the Issedones say merits but little attention." This skepticism by Herodotus came about those who claimed their cultural foundations to have been in Hyperboreans, when the Issedones asserted, "The ocean, they say, commencing at the east flows all around the earth; this they affirm without proving it." The concept that the earth was spherical first arose with the Hyperboreans long prior to Pythagoras, and was adopted by him. While the Issedones were the great civilization that once inhabited the Tarim Basin when it, the Gobi, was a flourishing fertile nation, their culture had migrated from north of the Celestial or Altai Mountains, home of Finns, Estonians Hungarians and numerous other Finno-Ugric tribes referred to as the Hyperboreans. Regarding these, whom Herodotus calls Argippei or Hungarians, Pliny and Mela had also described them as living in the forests and used berries largely for food. They were scanty of beard, and held inviolable by their neighbors, who left them alone, and their people took refuge with them undisturbed. Some of Alexander's historians highly commended the Abii (Ablai) believed to be another name for Argippei, for their justice and forbearance, for which they are celebrated far and wide. (See "Races of Man" by Calvin Kephart). Homer names certain Abian Scythians as the best and most just people on earth already in the 9th century B. C. Prior to the time of Herodotus it is said the Finns had migrated to the region described as Scythia between the east side of the Urals and Sayan Mts. In fact, in no way has any anthropologist ever been able to distinguish Nordic Scandinavians, Finns and other Baltic races except by different dialects of language.

Manuscripts of the Szeklers, Magyar mountaineers of Transylvania, who are the purest representatives of the race were written in mysterious old Runic characters, nearly identical to Ugrian writings found in Siberian caves. It runs from right to left, like Aramaic of the Essenes all relating to Pre-Sumerian, Pre-Minoan, etc. cultures. Hungarians are said to turn from Christianity to worship Hadur, a stern god whose temple is the whole outdoors and who demand sacrifice with righteous living, but is scornful of dogmas and formalisms; he is celebrated by assembly in the forest, where grains and flowers are tossed into flames to honor him. The Chinese have cultural origins they claim were obtained from the Uyghurs, Oigurs, or these Ugrian peoples, also called Eluths. Genghis Khan and his successors employed them as secretaries, the Great Khan Tamerlane of the so-called Mongol empire being "tall, strong, fair-skinned, red-haired, broad-shouldered and muscular, being a descendant of Nordic Aryan Finno-Ugrian origins ruling-family." they patronized Buddhism as far west as Finns and Scandinavian Lapps, and over-ran the Christian's Holy Land causing all of Europe to tremble and arm in the "Holy Crusades", all because of ignorance defending the Buddha who taught abstinence from bloodshed and hate, just as Christ, who upheld the God of Love.

Now, after these brief notes of mundane commentators, let me quote "On the Hyperboreans", a book by Hecataeus of Abdera of 4th century B.C. (Ancilla to Pre-Socratic Philosophers by Kathleen Freeman): "Elixoia island of the Hyperboreans, lies beyond the river Karambyka; the islanders are named Karambykas from the river. The Northern Ocean (Arctic) they call Amalchius from the River Parapanisus onward, which river irrigates Scythia, the name Amalchius in Scythian means 'frozen'. The Hyperboreans worship Apollo, the Sun God. These Hyperboreans survived to this day, worshipping Apollo who has been seen visiting them. There are three branches of Hyperboreans. In the place opposite the Celtic land on the Northern (Arctic) Ocean, is an island not smaller than Sicily; this belongs to the North (Prehistoric Hyperborea), and it is inhabited by the so-called Hyperboreans. It is fertile and productive, and of fine climate, and has two crops a year. They say the moon, being a short distance away, can be completely seen, and has several earthy projections clearly visible. It is said that Apollo visits the island every 19th year, and this is called by the Greeks a Great year, when all the constellations are completing their journeys. On his appearance the god plays the harp and dances all night long, from the spring equinox to the rise of the Pleiades, enjoying his own fine weather. This State is ruled by the Boreades, who are the descendants of Boreas, and the offices are hereditary." (The verbs are mostly translated in the present tense, since the interpreter was unaware that this speaks of a tiny fragment of ages-ago continent of Hyperborea that these descendants describe, and of formerly tropical vegetation with two crops a year. The Greek 19th year is our 20th year. The Arctic still has a Kara Sea.

There is much to tell about my Hyperborean kinfolk, and my early life on earth when I was better known as "Helios, the Hyperborean", a 100 thousand years ago. My reader's suspicion as to my identity therein may have already arisen in Volume I of my "MAITREYA" Autobiography on the first page of Chapter III wherein I give a diagram with a triangle of locations wherein I would recover the continuity of Godhood in Super-Consciousness, thru a "Period of Initiation for the Son of the Sun," Temple of Metta-Aum, etc. which I intuitively received early in 1940. With it there was a very significant anecdote, but the reasons for it were to be understood much later than I realized then: It is the Spiritual Archetype of Saviors and Sacred Scriptures of all Mankind's Religions. Quoting it reads: "Each day the Son of the Sun helped his Father the Sun drive this Golden Zephyr Chariot around the skies above the clouds. One day the Sun let his Son drive the Chariot alone. The Son had may temptatious friends and the Chariot became so heavy that it came too close to the earth and the fiery horses burned and set fire to the earth. Ever afterwards the Sun did not permit his Son to drive the Chariot alone, and allowed only those friends who benefitted his Son the most." This intuitive diagram and notes came along with a clairvoyant vision in which my body became purer and purer, until it was so pure that it rose in the air resembling the Dharma Megha (cloud of virtue) witnessed in the Highest States of Samadhi, unable to remain in the heavy dense environment of fellow fruitarians who had no higher goals than the sensuous enjoyment of tropical fruits, an idle life without work in a warm climate. This did symbolically come about in that for health alone, I had to abandon the tropical equatorial lowlands of those 1940 companions, for the highlands, eventually living by Lake Quilotoa in a volcanic crater, beside various other occasions in various situations. In time, my spiritual mission had a cosmic meaning.

As to the SUN or HELIOS in Greek, it historically also became the Greek God "Helios", and identified with Phoebus Apollo. Reference book data show "Helios" to be "the Sun god, son of

Hyperion and Theia. Each day he drove this golden chariot from the palace in the East to his palace in the West. he symbolized the material aspects of the son, being described as Apollo in Spiritual aspects. In the Odyssey of Homer,- Helios shipwrecked Odyssey for eating his Sacred Cattle. He is called Sol by the Romans." (Columbia Viking Encycl.) The same source describes "PHOEBUS APOLLO" as "the Olympian God of Light, Music, Poetry, Pastoral Pursuits and Prophecy. As a god of music the lyre was sacred to him, and he was the father of Orpheus (source of many Pythagorean ideas). He was also the god of Healing, and the father of Asclepius (the first healer and Early teacher of Gnostic Philosophy and symbolism), He is identified with Helios, the Son God. He was the Son of Zeus and Leto, and worshipped at the Temple of Delphi as the God of Prophecy."

In the "HEALING GODSPELL OF ST. JOHN", we describe the practice of Natural Sun worship, or "Heliotherapy". Therein one may contemplate the light colored clouds which the dawn leads out into the heavenly fields to graze like the cattle of Helios. Praise Helios, while the Hyperborean sons of the Sun toil to bring forth the Fruits of their Gardens, bodily empowered by the strength of Hercules. Then in the Glory of the Father, Life-Giver, they commune partaking of Juicy Fruits filled with Golden Nectar of Gods, the Living Water of Life Everlasting. In the crimson beauty of sunset, the western skies paint the final celestial CROSSing of the Lord of Life, as all his children lay down their bodies in sacrifice during the night, and again welcome their Father, as their hearts cry joyously "HE HAS RISEN", in the Cosmic Dawn's Embrace with arms outstretched as they meet their Sustenance and Savior.

When I wrote down the "Period of Initiation" verse early in 1940, living in a palm leaf tree abode 16 feet above the ground in the Florida Everglades, on a juicy fruit diet, basking long hours in the sun daily, I had not read a page from the Christian Bible of our time, being more aware of the Spiritual Allegory and Symbolism which firsthand I had participated of in Life as Helios, Krishna, Buddha, Christ, etc.

Lecturing on the Science of Religion, Prof. Max Muller described the significance delightfully: "One of the earliest objects that would strike and stir the mind of man, and for which a sign or name would soon be wanted, is surely the SUN. It is very hard for us to realize the feelings with which the first dwellers on earth looked upon the Sun, or to under-stand fully what they meant by morning prayer or a morning sacrifice. But think of man at the dawn of time, think of the Sun awakening the eyes of man from sleep, and his mind from slumber! Was not the sunrise to him the first wonder, the first beginning of all reflection, all thought, all philosophy? Was it not to him the first revelation, the first beginning of all Trust, of all Religion?".

"We must not be surprised at finding, on close examination," wrote Sir Wm. Jones (Asiatic Researches), "that the characters of all the Pagan deities, male and female, melt into each other, and at last into one or two; for it seems as well founded opinion, that the whole crowd of gods and goddesses of ancient Rome, and modern Varanes, mean only the powers of nature, and principally those of the SUN, expressed in variety of ways and by a multitude of fanciful names." What the "Pagan" religions did, Christianity did better, with a great many sources to synthesize its Spiritual Allegories, Symbols and Artistic Beauty.

"Christ Jesus, then, is the SUN, in his short career and early death. He is the child of the Dawn, whose soft violet hues tint the clouds of early morn; his Father being the Sky, the HEAVENLY FATHER, who has looked down with love upon the Dawn, and overshadowed her. When his career on earth is ended, and he expires, the loving mother, who parted from him in the morning of Life, is at his side, looking on the death of the Son whom she cannot save from doom which falls on him, while her tears fall on his body like rain at sundown. From her he is parted at the beginning of his course, to her he is united at its close. But Christ Jesus, like Chrishna, Buddha, Osiris, Horus, Mithras, Apollo, Atys and others, RISES AGAIN...when the night is done." (Bible Myths by T. W. Doane). The early Fathers of the Christian Church admitted that: "those who live according to the Logos were really Christians...among the Greeks" (St. Clement of Alexandria); "the Christian Religion is neither new nor strange, but was known to the ancients since the beginning of the human race." (both St Augustine and Eusebius) "Our master Jesus Christ, to be born of a virgin without any human mixture, to be crucified and dead, and to have rose again, and ascended to heaven, we say no more in this than what you say of those Sons of Jove" (Justin Martyr to Emperor Adrian,) and we could quote dozens of authorities, the same thesis, that "Nothing distinguishes you from the Pagans" which as both Faustus, writing to St. Augustine, and Ammonius Saccus, the Greek have illustrated. No one here argued that these named Saviors, Sages and Saints have not existed as humans around whom these solar symbolisms, allegories and mysterious happenings have been worked in.

Plato in the fifth century B.C. philosophized: "The Power of Supreme God (in the Son of God) next to Him was decussated or figured in the shape of the Cross on the Universe." Jesus Christ was called the Rose of Sharon, of Isuren, and the Nautrutz or Natsir of the Nazarene is symbolized with a Rose blossoming on a Cross symbolizing Spiritual Unfoldment. However, the Crucified Sun was known since the Hyperborean Helios emanating loving warmth and light unto mankind. Brahmin, Taoist and Buddhist records of the Illumined Sages of old extend hundreds of thousands of years into the past so the next chapters will illustrate their wisdom and cultural beauty as a model for the return of Satya Yuga.

CHAPTER II: THE ANCIENT EASTERN LEGENDS OF PARADISIAN ORIGINS

To arrive at a Universal Conception of the Beginning and End of Mankind, the Genesis and the Apocalyptical Last days as we describe them in the West, we must re-integrate both a Universal Western Doctrine and a Universal Eastern Doctrine. In earlier writings about the Vitalogical Sciences, I have fairly covered the Western, that is, a doctrine that Esoteric Students of Gnostic Christianity would understand, seeing that the Catholic, Protestant and other New Testament Christians, the Moslem or Islamic Sufis, and the Hebraic Bible followers, use the same basic Scriptures known as the Genesis. They all claim to be heirs to Abraham. But the word Abraham or Abram in the original shorter form, whose wife is Sarai or Sarah, is related to a Universal Brahm (Brahma) and Sarasvati (Sri, wife of the Moon), and in the SECRET DOCTRINE, Mdme. Blavatsky describes as concurring with an eclipse of the SUN (Brahm or A-Brahm) at the ascending node of the MOON (associated with Sarasvati, or Sri, wife of Brahm, divided from his Being as part, a Goddess of Wisdom, Gnosis and Speech). Altho the legends of Brahm and A-Brahm are similar in significance, the prefix "A" before Brahm denotes without or Void of Brahm, that is coming before or Supreme. Abram means the Father is the Highest, so that in the Legends of the Genesis we can well expect that the Western and Eastern Doctrines are parts or halves of one original Legend, that became divided into Semetic and Aryan, or Eastern and Western versions.

The Eastern version is to be found in the "DIALOGUES OF THE BUDDHA" or Part III, "The Aggana Suttanta" described as the "THE PASSING AWAY AND BECOMING OF THE WORLD", since the Genesis and End are related in a wheel of Birth and Death. Before I begin quoting, let me clarify my reason for choosing a Buddhist Scripture, as a Universal Eastern theme, in that the Vajrayana (Middle Path) or Southern doctrine, Brahmanism, Taoism, or Eastern Doctrine in General, and in a basic sense, even find parallels or coincide with Western Doctrine in the end. The scripture speaks for itself:

The Exalted One (the Buddha) said to Vasettha: "There comes

a time when, sooner or later, after a lapse of a long, long period, this world passes away. And when this happened, beings have mostly been born in the World of the Radiance; and there they dwell, made of Mind, feeding in Rapture, Self-Luminous, traversing the air, continuing in their Glory; and thus they remain for a long, long time.

There comes also a time, Vasettha, when sooner or later this world begins to Re-Evolve! When this happens, beings who have deceased from the (Astral) World of Radiance usually come to life as humans. And they become made of Mind, feeding on Rapture, self luminous, traversing the air, continuing in glory, and remain thus for a long, long time.

Now at that time, all had become one world of water, dark and of Darkness that maketh blind. Neither moon, nor sun appeared, no stars nor constellations, neither night nor day manifest, no months nor years, and neither female nor male. Beings were reckoned just as beings only: And for those beings, Vasettha, sooner or later after a long time, the earth was spread out in the water. Even as a scum forms on the surface of boiled milky rice that is cooling, so did the earth appear. It became endowed with color, with odor and with taste. And in places (under the trees of golden fruits, mangos, apples, peaches, apricots, etc.), it appeared like well made ghee or pure butter, so was its color; even as the flawless honey of the bee, so sweet was this fruit of the earth.

Then Vasettha, some greedy dispositioned being said, "Lo what is this? And tasting of this savory morsel or product of earth, craving entered this being. Then, others followed in tasting and consequently feasting of this delicious golden fruit that earth yielded. But alas, as they feasted of these golden lumps of earthy food, their own self luminance faded away, and for this lack, luminous bodies manifested only in the sun, moon, stars and constellations. Time, years, months, days and nights manifest. Once again, the world evolved as it has. And to the measure that beings ate of earth-born food, their BODIES BECAME SOLID, with variety in their comeliness. Like the well flavored and ill flavor fruits born by earth, beings became well-attributed and of ill-attributes. So, those endowed with beauty, became vain and conceited, and those ill-attributed, despised. Thus, as beings became well and ill attributed, the golden well flavored fruits of earth disappeared, and men lamented the loss of qualities they experienced in food, having appropriated such attributes for themselves."

It is evident that the fat rice-fed Buddhist monk who recopied these scriptures, like retold legends, probably never knew the juicy golden fruits mentioned in the Ancient Genesis, nor was first man able to describe golden fruits covering earth in the Garden of Paradise, not having the necessary words or associations invented, so the illogical term "lumps of savory earth". The only comparison in such monk's minds is thus eloquently spoken by descriptions saying it was like the milky rice pudding, with butter or ghee and honey, that filled their minds before making daily rounds to receive alms offerings in their bowls.

Briefly, to condense another page of descriptions, when the above mentioned food disappeared, the soil produced lowly outgrowths like mushrooms in size, also of good taste and color obviously referring to vegetables or plant leaves. However, the monk's minds only can imagine and describe it as like mushrooms, ghee and honey, so familiar to them. Human minds became denser in thinking and bodies denser in earthy substance. Then, as told in Hebraic Genesis 8:17,- after the earthy plant life lost its attributes,- creeping things appeared on earth, and man ate of them. Their bodies and minds became very dense, they became even more conceited, and among themselves, despised one another, and thus even creeping things disappeared, without mankind understanding the significance thereof. Now here comes the part well-worth quoting:

"Then, Vasettha, when the creeping things had vanished for those beings, RICE APPEARED RIPENING IN OPEN CLEARINGS. Then those beings feasting on this rice in the clearings, nourished by it, continued for a long while. And in the measure as they fed, so did their bodies wax more solid, divergent from very gross to comely in pronounced features. In the (grain-fed) female the pronounced features of a female, and in the male, the manly features. Truly did woman contemplate man over-much and man woman. And consequently PASSION arose, and burning desires entered their flesh, causing them to engage in the satisfaction of the lusts of their flesh. And beings seeing them so doing, threw sand, ashes, cow dung, (now rice, confetti and old shoes at the bride and groom) crying: Perish, foul one! Perish, evil one! How can a being treat another being so? Even so when the groom leads the bride away by their ancient tradition, people throw Ashes, sand, or cowdung at them, not knowing the ancient significance thereof."

This Buddhist text matches the description given in the Hebraic Genesis in significance, as we have taught in the Vitalogical Sciences. Eating seeds, grains (from which bread was made) brought on sexual lust opening their eyes, so they become so ashamed they must wear fig-aprons to hide their organs, and everyone laments the original sin that humans suffer to the utmost extreme to this day, unable to control the population explosion, and lack of food and things for mankind along with man's polluting of one another to survive in body and mind. At the end of each incident, the Buddha wisely warned that man had not understood the significance thereof, and thus our Vitalogical Mission doing this.

Then, the Buddha explained how, gathering more and more rice, so as to last for two days, one week, a month and so on, the hoarding of rice began to occupy man's body and mind, and in the same manner the quality obtainable became worse. Today in Ecuador, even, there is rice hoarding every year, prices soar, famine develops, and penal laws punish whoever is discovered hoarding grain. Thus, the Buddha identically illustrates how men became greedier and greedier with rice, and then fenced off plots against neighbors and invaders, and consequent attacking of one another for violations. With stealing, lying, committing violence and such crimes, it became necessary to select a Lord of the fields (Maha Sammata) or Great Elect as ruler along with other royal noblemen, who would censure evil deeds. Thus, the Buddha explains to Vasettha, how the upper cast of Brahmanas came about alongside those who were considered outcasts by tradition. He tells how capital punishment came to be used seeking to prevent crime, anointed kings governed, weapons and armies were developed, and from former sublime mind beings (or Gods) with a life-span of 80 thousand years, the lifespan is reduced to half, then progressively down to just one century. As we have described in other writings, everything is disintegrating, passing away and such is the fate of civilized, worldly beings bound to pleasures.

Living in this flesh-bound world, materialistic civilization and the earth's modern chaos, my Spirit especially longs to return to our original "SHAMBHALA", the Divinely Blessed Isle of Paradise of the old Greek legends of Hyperborea, with its "Heavenly Mountains" (or Altai Mts. of today), located in a vast inland Sea of Wisdom (but now the Gobi), from where the Dhyan Chohans, Sons of God or Elohim, have guided the destiny of this planet, earth, thru countless disasters and cataclysms, to this climatic episode speaking forth the Birth of a New Golden Age (Satya Yuga). This is the work of the "AUGUST ORDER OF LIVING IMMORTALS IN CONSCIENTIOUS TATTWAS" (the High Heavenly Hierarchy of the Paradisians), that we herein record since the ancient-most references of those events which our readers would seek authorization beside my own memoirs. In the days of my eternal youth, roundly a 100,000 years ago, I remember, as ever, passing carnal birth, conceived and born from a cave underneath my mother's heart also, as well as the sweat and egg born existences. But since that period our race of Immortals, having become sense-bound mortals conscious only of their immediate times, have found it necessary to leave visible records, which Anthropology, Ethnology, Philology, etc. has finally stumbled upon the evidence of.

"It is maintained by learned men that geographically man began his career, not in Syria, but in what is now a barren wilderness on the northern slope of the Hindu Kush. The `Garden' therefore stretched to Taurus (S. Turkey) and the (Arctic) Polar Sea, eastward to the Altai, or the Celestial Mountains of the Chinese and its western borders were Ararat and Caucasus. Geological research has proven that a vast inland sea existed aforetime, which filled the whole area between the site of Constantinople on the West, and Turkistan on the East, a length of 2,000 miles by about 1,000 miles broad. The Caspian of Azoff and Aral Seas are but ponds left when this great area was drained. The Hebraic legend of Genesis, the Bacterian one in Vendidad (Zoroastrian), and at least a score of others so strongly confirm the same story, that it is reasonable to believe it has a foundation in truth." So has the quoted researcher, Dr. Paschal B. Randolph concluded in his detailed work, "PRE-ADAMITE MAN: Existence of Human Race upon earth 100,000 years ago".

Scientific findings establish that "Mongolia has not been submerged since the Age of the Reptiles, about 175,000 years ago. Central Asia embraces the largest area of continuously dry land known to science", states Calvin Kephart (Races of Mankind). He referred to this period as the Riss-Wurm Inter-glacial Period, or before the Wurm glaciation (Ice Age), and Mongolia is now mostly Gobi or Shamo Desert, south of Altai or the "Heavenly Mts." of Chinese legends.

Now, to better understand of what I speak as I reminisce of Shambhala, I should define the descriptive terms used. "Shama" in Sanskrit means Peace, Serenity or Contentment, rooting from the Hyperborean word "Sama" which means the same, just the same as ever without extremes or opposites, of similar equivalent universally. To express oneness with God, the Brahmins call it "Sam-Adhi" (Samadhi), while the Buddhists speak of "Sam-Bodhi" (Sambodi), or Oneness with Truth or Supreme Wisdom, referring to Dhyana Chohans or Buddhas countless ages before Gautama. "Bha" is a syllable meaning Light, Splendor, Beauty or Illumination, while "La" is a suffix denoting a site or status. So in combination we have "Sham-Bha-La with lofty interpretations as the Blessedness or Bliss of Heavenly Paradise on earth (like New Uru-Salem, Jerusalem of John) that Yogis, Taoists, Sufis, etc. like the Buddhists, seek to obtain for Omniscient Oneness of God of Illumination. Such a place existed (and now exists embryonic-ally) in Satya Yuga or the Golden Age, Satya being Supreme Truth, the heavenly mansion of Brahma, the Paradise of the Sons of God who were able to face and see the Light Supreme.

However, today, in the battle waged by the new cult of "Modern Science" which seeks to stifle the last recollections of Paradise as well as equivalent Eastern doctrines, the cult of the "Shamans" has now been under attack. From Shaman, once a term for mendicant monks who shaped religious orders universally,- sarcasms have developed such as a "sham" denoting a trick, have proceeded. Webster's dictionary traces Shaman to the Ural Mts. and the Altai (Celestial) Mts., region of these Physician-Priests of earliest traditions of mankind. In the Sun-Worship of Assyrians, "Shamash" means the Sun, the "Mazda" of Persians. These Sun-Worshippers, or Shamans were believed to be the "Gymnosophists" or earliest origin of the Essenes, beside the Bhon or Red Cap Lamas of Tibet, also mentioned as Gautama Buddha's teacher, as well as earliest Hindu legends. Bible translator, Geo. Lamsa, points to the Sabean and Essene doctrine of Gnostic Christians of St. John, who abstained from wine, meat and certain foods, brought over from the "Samash"(Sun) Worshippers told of in IV Kings 17 of the Bible. Herodotus gave all kinds of tales about the wisdom of the Hyperboreans affirmed by the Issedones in that not only did they claim the earth to be a globe, but he quotes Aristeas who told of being inspired by Apollo to go to the land of the Issedones, and beyond them dwelled the Arimaspians, a people that have only one eye...When one has passed thru a considerable extent of the rugged country (of the Scythians), a people are found living at the foot of lofty mountains (the Altai or Heavenly Mts.) who are said to be bald from birth, both men and women alike, and they are flat nose and have large chins... "These bald people say, what to me is incredible, that even people with goats feet inhabit these mountains,; when one has passed beyond them, other men are found, who sleep six months at a time..." Herodotus then tells us also that the Issedones were said to celebrate the sacrificial lamb festivity when a father died, in which both man and sheep are cut up, well sliced and mingled, served and eaten. The head of this man is preserved as a sacred image. "In other respects the Issedones appear to have sound enough sense of the difference between right and wrong, and a remarkable thing about them is that men and women have equal authority."

This is a good test to find out who the true disciples of the Esoteric Vitalogical Sciences are, or separate the "sheep and the goats", since those only seeing material and physical ends in natural living, moral vegetarians, health cultists, etc. will disbelieve or refuse to subscribe to any such findings in Mystical Anthropology. The Mystic probably already has a rudimentary development of the third eye of the early one-eyed giants among my people, which has all but atrophied completely in the pineal body of the present civilized race. The Hyperborean giants with only one eye in the middle of their forehead, had not partaken of the Tree of the Knowledge of Good and Evil by habit, and to live eating from both the seed or nut Trees, beside the Trees yielding juicy fruits, indiscriminatingly, is without this basic Living Water Alchemy, which opened up the present double-eyed vision. With a perfection of the race, there remains little need of perspective, or a looking for two sides to everything, good and evil, left and right, sin and sanctity. As soon as the former Paradisians were cast out of their Paradise with the changing of the earth's axis, and the Polar Lands turned from a warm sunny climate to the present frozen Arctic tundra, those who survived mainly gathered to the Tarim and Gobi regions. It was there that our present civilized way of life was developed and our characteristic racial qualities developed. In the process, only the freak monster babies produced by certain drugs used during pregnancy can describe the incredible racial deformities.

Take for instance the fact which has created much Biblic debate,- "God created man in his own image, in the image of God he created him; male and female he created them", in which many hold that before Adam sinned (eating seeds which brought on sexual differences) mankind consisted of true Sons of God, both male and female in virtues. This anyone can visibly see if one observes how males have rudimentary mammary nipples, now atrophied and without use, altho Adam supposedly gave birth to his flesh and bone of body in Eve, and therefore must have nourished her thru functioning mammary glands. Furthermore, males began to grow beards even on the face, just as both male and female not only developed vulnerable erotic areas of the body, today being vulnerable to cancer, syphilis, etc. beside related psycho-pathologies, while hair also developed in these sexual areas. Hair is known to be a good index as to poisons used in homicide and with the change from living water of juicy fruits, to water, grains, etc. heavily laden with calcareous substances, the erotic areas sought to eliminate them, as do the seeds of plants. My mother, whose genetic body I perpetuate, in her childhood often contemplated the sunrise over those Altai or Heavenly Mountains, and could assure that at our genetic source in Helios, the Great Hyperborea, that those goat-footed mt. tribes had to be as surefooted as described to survive free, unbounded, as the plains-folk became by civilized customs. As to sleeping six months, this has often inspired me, especially during the 6 month fasts, wherein I was imitating the bears who only joined us in the apple harvests, spending the rest of the year in suspended animation, like practiced by Yogis, etc. in contemplation.

Finally, as to eating the sacrificial lamb or Pasch mingled in species with the Father's flesh by the Sons of the Sun, this I well do know, that such is ritual in Christian Churches. "You know me: then you also know the Heavenly Father!" Jesus said, "Unless you eat of the flesh of the son of man, and drink his blood, you shall not have life in you." In the Holy Communion, bread is believed to be the flesh of the Son of man and God, and wine, his blood. This is the Living Bread that came down from Heaven. Jesus is also the Supreme Sacrificial lamb that redeemed mankind from sins. This is why in Paradise it says, "God said, I have given you every plant yielding seed, and every tree that bears fruit yielding seed: to you it shall be meat." Any food can be used to have flesh on our bodies, so in simple Hyperborean language the two are same: juicy fruits naturally yield the most living flesh and pure red blood. Grains were first bred from seed bearing grasses in the period slightly before 10,000 and no more than 15,000 years ago, mostly in the Issedon inhabited Gobi-Tarim area, destroying the wooded forest to raise grain, accounting for their ecological desolation, cities now buried in wind-blown sands of the desert. No sooner had they raised Cain (the Biblic grain-grower) here, than racial tribes divided by their genetic deformities, immigrated to Sumeria, Egypt, India, Arabia, etc. with the same consequent results. All of this is mystically told in descriptions of the unbloody sacrifice of bread and the fruit of the vine in the Gnostic-Essene tradition, and the profane mocking of rituals in actual crucifixions centuries and millenniums before Jesus Christ. Herodotus of the above account dates 484 to 425 B.C., and the history of the Assyrian king, Assur-nazir-pal records boasting cruelty for vast conquests, tells of hundreds of CRUCIFIXIONS among resisting conquests, 700 being crucified in Bit-Ura in land of Dirra the NINTH CENTURY B.C. The Gospel of Jesus Christ existed from the start of this world civilization, shadowed in the rituals and traditions of nations. At the beginning of Kali Yuga, this past Dark Age, which was said to have started after Krishna 5,000 years ago (3,000 B.C.), we have the earliest detailed description of a Savior and his crucifixion and resulting ritual of which Christianity was a progressive amelioration in legends, and thus provided the ready explanation as to the Herodotus, and Assurnazirpal rites of sacrificial lamb and crucifixions, described above. At the foot of the Heavenly Mts., north of the Himalayas, called Mt. Meru by the Hindus, under the guidance of Vasichta with a group of devout virgins, Devaka gave birth to Krishna, Edward Shure tells us in "The Ancient Mysteries of the East". Vasichta had lived 60 years on a diet of wild fruits only, and in this Paradisian setting among arbors of fruit trees whose branches planted themselves in the soil again in infinite arcades, the child Krishna grew strong and filled with Wisdom. This should account for why the Christian Bible dates from 5,000 years ago, time of Krishna, and roots in A-Braham in the tales of Patriarchs.

Of even more factual findings have been my researches into Chinese Shaman legend of Taoist Alchemy. In the "Secret Life of Plants", Tomkins and Bird state that recently the whole scientific world was shaken by facts on how living things transmute one element into another. They quote from many sources of proof, and leave the science of biochemistry even in bewildering cobwebs of theory, since the Living Water in plants is able to transmute one element of the soil into another of plant life, and humans eating plant life receive still other elements from those contained in the plants eaten. This evasive Alchemy of Biological Transmutations became an early grounds for research by the Taoist Wise men, and some of the earliest records in principles of our teachings.

"Far away on Mt. Ku (also described as Magical Isle of the Blest) lives the Spiritualized (Heavenly or Holy) man whose skin is white as snow and he is gentle as a young girl. He does not eat any of the 5 grains, but inhales the air and drinks the dew. He rides on clouds and can mount the flying dragons to wander beyond the four seas. By using his spiritual powers, he can protect us from sickness and decay, and ensure a rich harvest," wrote Chuang Tzu. P'eng lai, is also described as Pu-An-Ku or P'eng of Ku, when describing the Isle of the Blest where the "Hsien" (Immortal Sages) live. This primeval man came out of the mundane egg and lived for 18,000 years, and the Chinese speak of events that transpired among their ancestors 129,600 years ago, wrote Dr. P.B. Randolph.

The Interior Gods Taoist School of Hygiene considered the divinities (or angels) as the necessities of life. The followers of this School avoid the three Worms of disease, old age and death, by not eating any of the 5 grains. They exist on wild herbs and fruit, and some get down to nothing but jujubes. The advanced adept eats no solid food at all, because solid food fills the body with excrement, and excrement inhibits the circulation of breath. To open up the breath channels the adept practices gymnastics, Tao yin, stretching like a monkey, and twisting his neck like an owl. Then he practices Embryonic Breathing like a child in his mother's womb. The breath is held 120 heart beats, and when he can do 1,000 he approaches Immortality. Holding his breath he practices interior vision, seeing within the body. He corks "his breath, while his tongue is raised to the roof of the mouth, and makes breath and saliva his pure nourishment. His body becomes light and he casts no shadow being transparent. Uniting the breath with the semen, he forms the Mysterious Embryo, and nourishing it with breath he develops a new pure body inside the old. Shedding the old he is Immortal. This paragraph is summarized from Holmes Welch's book on "Taoism", and also I wrote about it in our "Transcendent Truth Teachings" in 1974.

In the above quotation about the Isle of Ku, which later became Mt. Ku when inland sea dried up, home of Heavenly man of the Heavenly Mts., there were other Isles that became mountains, such as Mt. Ku-un-lun (north of Lhasa, Tibet), where in turn the Heavenly Goddess, Hsi Weng Mu lived on the Peaches of Immortality. Kuu in Hyperborean means, "moon", while in Sanskrit "Kuhu" is the new moon goddess, and in Vedic writing is "the moon" in meaning. Thus, it is the name of the Nadi (or channel of vital nerve or energy current) that leads to the generative organs. Mu in Sanskrit means bonds, ties, and in Hyperborean usage it comes in "muda" meaning mire, mud or slime, "muld" or earth, and "mure" or trouble, worry, while "muu" is something else. Due to my studies with "The Lemurian Fellowship" starting in 1936, "Mu" was ever given as another word for "Motherland", but in the inspired ideals the Fellowship promoted, was the concept that "the mud flats" of Lemuria are now submerged under the Pacific Ocean, from where the survivors colonized the Great Uighur Empire in the Gobi which then was fertile country. But this, in time, I found to be a twisted, biast opinion, and while some eastern writings refer to its origin in the sunrise sea, this was also of the northern sea after the axis change then the equator of Hyperborea came to transverse the mid-Pacific, as it now does. Thus, they had confused "Lemuria" with Hyperborea, and later "Atlantis" in reality.

Now, for the unaltered concept, before the converted confusion of writers of this century, let us go to our 1906 "New Standard Encyclopedia" (University Society) for a concept of ancient times until our time. LEMURIA: a hypothetical continent supposed by some to have at one time extended from Madagascar and South Africa across what now is the Indian Ocean to the Asiatic Archipelago (East Indies), named from its corresponding with habitat of lemures. LEMURES: the genial designation given by Romans to all the spirits of the departed persons, of whom the good were designated Lares, and the bad, Larvae, and were feared as capable in their night journeys of exerting a malignant influence upon mortals. The festival called Lemuria was held the 9th, 11th and 13th of May, and was accompanied by ceremonies of washing the hands, throwing black beans over the head, etc. and pronouncing 9 times these words, "Be gone you specters of the house", which deprived the Lemures of their power to harm. Ovid describes the Lemuria in his Fifth Book of his "Fasti". ATLANTIS Atlantica: an island said by Plato and others to have once existed in the ocean immediately beyond the straits of Glades, that is in what now is called the Atlantic Ocean, a short distance west of the Straits of Gibraltar. Homer, Horace and some others made two Atlanticas distinguished as Hesperidia and Elysian Fields, and believed to be the abode of the blessed. Plato states that an easy passage existed from one Atlantis into the other islands which lay near a continent exceeding in size all of Europe and Asia. Some have thought this to be America. Atlantis is represented as having sunk beneath the waves, leaving only rocks, shoals in its place. Geologists have discovered that the coast line of Western Europe did once run farther in direction of America than now, but the submergence seems to have taken place long before historic times, so that the whole ancient story of Atlantis was founded on erroneous information or arose from a clever guess put forth by a man of a lively imagination. HYPERBOREANS: dwellers beyond Boreas or the North Wind, a name given by ancients to a mythical people, whose land was generally supposed to lie in the extreme northern part of the world. As a favorite of Apollo, they enjoyed an earthy Paradise, a bright sky, a perpetual spring, a fruitful land, unbroken peace and everlasting youth.

Identical locations were given in "THE SECRET DOCTRINE" of Mdme. H.P. Blavatsky in 1888 as to the locations of the three prehistoric continents, and these are classified by location as the homes of the 7 Root Races. 1. Sacred Land, home of First Root Race that has no beginning or end, also had no earthy existent location like the astral plane. 2. Hyperborean continent, home of Second Root Race is of the northern part of the globe; 3. Lemurian continent, home of the Third Root Race is of what now is the Indian Ocean, Africa, East Indies and south India, with black, pygmy to giant peoples. 4. Atlantian continent, home of the Fourth Root Race is the origin of the Redskin Indians of the Americas, the Yellow Races of China and Japan. 5. The Aryan Lands, original home of the Aryan or fifth Root Race involving Europe and Asia. 6. North American continent, or home of the Sixth Root Race. 7. South American continent, is to be the home of the Seventh Root Race after the other mentioned Lands and Races fade into the past. Each Root Race had its sub-races, or newborn divisions, so that the complete Racial Type will have attained to Seven Planes of Integrated Consciousness, Superconsciousness. In "Isis Unveiled", Mdme. Blavatsky explains that "Even the oldest of sciences, their Kabalistic 'secret doctrine", may be traced in each detail to its primeval source, Upper India, or Turkestan, far before the time of the distinct separation between Aryan and Semitic nations. The King Solomon so celebrated by posterity, as Josephus says, for his magical skills, got his secret learning from India, thru Hiram, the king of Ophir, and perhaps Sheba. (India and Arabia)

Thus, "Muu", like in the Hyperborean tongue, means something else or someone else, and not Lemuria to most of us. "Maa" in Hyperborean means soil, earth or land, and in Sanskrit water, season or good luck, while in Egyptian she was Mother Nature,

Goddess of the North Wind.

In an out-of-print book I published during my 1954, 6 month 17 day fast, on "VITARIAN COMMENTARIES", in relation to my Finnish Racial heritage, I described an identical description quoted above about the Hyperboreans, beside adding that they ate only fruit, but originally, like Gods, from whom they descended, they subsisted on air and sunshine or were "Helioarians" (Helio-Aryans) as also described in "Spiritualizing Dietetics-Vitarianism". In continuation I wrote, "Modern Anthropology uses Hyperborean to designate the people of northeastern Asiatic origins, who cannot be classified among the American Indians or Asiatic people. Hindus continually refer to Mt. Meru, the Heavenly Throne, or mountain of the Gods, North of the Himalayas. In an 1896 Expedition to Tibet, records were found to lead to the discovery of the capital of the Great Uighur Empire in the Gobi Desert, at Khara Kota. The Tibetan monasteries (recently contained) some of these Naascal writings that say, "the Naacals", 70,000 years ago, "brought to the Uighur capital cities, some of the Sacred Inspired Writings of the Motherland. Uighur was the most important colonial empire of the Empire of the Sun (Helios, Apollo). The history of the Uighurs is that of the Aryan races who formed chains of settlements across central Europe. The symbol of the + (cross), man's oldest symbol, geometrically locates a one dimensional point of being (consciousness or life), and is found in the 70,000 year old Naacal writing. The Tertiary Uighur Empire was destroyed by the Great Flood that destroyed the Motherland (Mu) and flowed over east Asia destroying the fruit forests of the fertile plains out of which the Himalayan and Ural mountains have heaved up, leaving only a desert, sand and gravel covering the ancient remains of the Ancient Empire about 20,000 years ago. Chinese records, 600 B.C., describe the Uighurs as having milk white skins, blue eyes and refined features. The Hindus claim their Aryan forefathers were long nosed, fair, tall and commandingly long haired and refined people who invaded India from the North. The "Homo Sapiens Nordicus" (Hyperborean) of the Finno-Uighrian branch of the race is of tall, flaxen haired, blue eyed Taevastan Finnic race of Estonia and southern Finland, contrast with the shorter blonde Lapps, who later populated most of the northern Scandinavian peninsula or Finland, Sweden and Norway. Taevastan means Heavenly in Hyperborean, presumably from Heavenly Man and Mts. (Altai Mts.).

The Hindu Legend of Paradise reads as follows: "In the Sacred Mountain MERU, which is perpetually clothed in the golden rays of the Sun, and whose lofty summit reaches into HEAVEN, no sinful man can exist. It is guarded by a dreadful dragon. It is adorned with many celestial plants and trees, and is watered by four rivers, which thence separate and flow to the four chief directions." In "Bible Myths", T. W. Doane also quotes Bunsen, "The records about the Tree of Life are the sublimest proofs of the unity and continuity of tradition, and of its Eastern origin. THE JUICE OF THE FRUIT OF THE SACRED TREE, like the tree itself, WAS CALLED 'SOMA' in Sanskrit, and 'HOOMA' in Zend; it was revered as the Life Preserving Essence."...."A drawing, brought by Colonel Coombs from a sculptured column in a cave temple of South India, represent the first pair at the foot of the ambrosial (Soma) tree, and a Serpent entwined among the heavily laden boughs, - presenting to them some of the fruit from his mouth." Doane then describes how ancient Hindu religion converts this into Phallic symbolism, like many have tried to convert the meaning of Tree of Life in Hebraic Scripture: "Siva, the Supreme Being, desired to tempt Brahma (who had taken human form, and was called Swayambhura,- Son of the Self-Existent), and from this object he dropped from heaven a blossom of the sacred fig tree. Swayambhura, instigated by his wife, Satarupa, endeavors to obtain this blossom, thinking its possession will render him Immortal and Divine; but when he has succeeded in doing so, he is cursed by Siva, and doomed to misery and degradation. The Sacred Indian fig is endowed by Brahmins and Buddhists with mysterious significance, as the Tree of Knowledge or "Intelligence". This is the Bodhi Tree of Buddha.

However, the species "Ficus Religiosa" of the Bodhi tree of Illumination, like the bad figs of the N.T. Bible, does not produce edible fruits with sweet Soma or juice. St. John, Mary and Martha are of the "Abode of Good Figs", the meaning of Bethany. Hindu Yogis, like Chinese legends blame woman and sex,- because they live on grains and seeds,- for all man's evils, but if they lived on juicy figs, no evil legends.

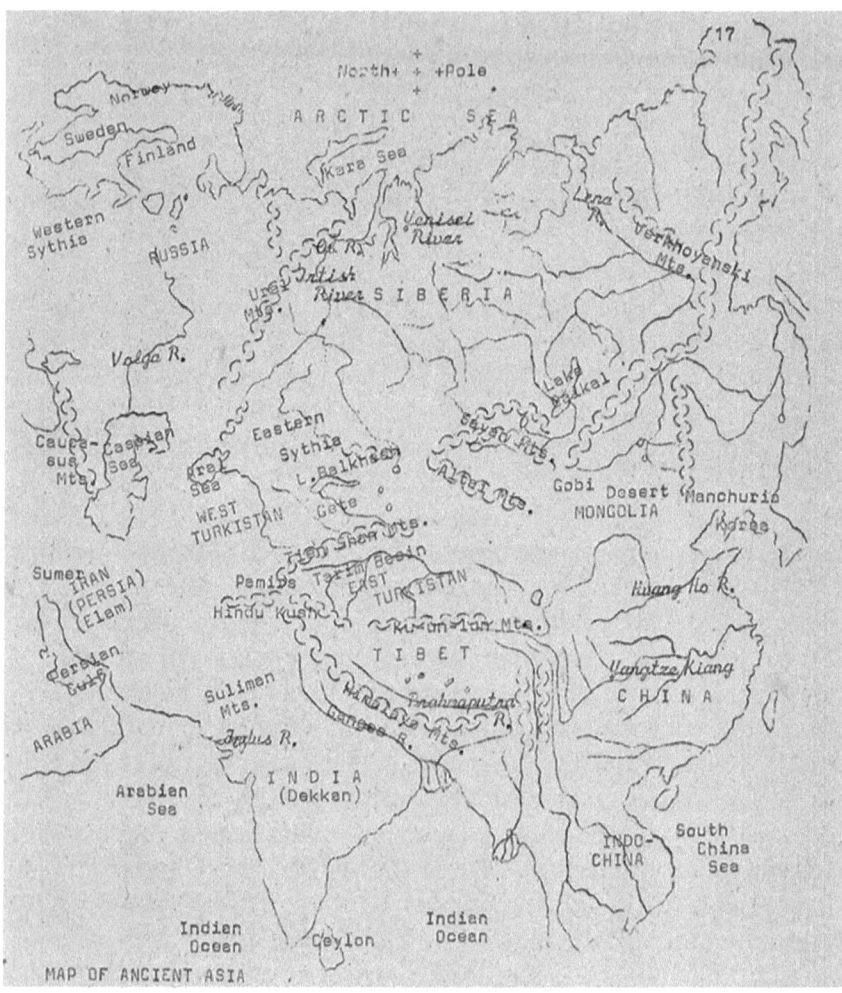

MAP OF ANCIENT ASIA

We have located some of the most important Ancient land Marks so the reader of our text will not be confused, altho a much better view can be obtained from maps, texts on history and geography, etc. We speak of the Hyperborean origins to be found in the Irtish and Yenisei River basin South of Kara Sea, Lake Baikal, and the Altai or Heavenly Mt. Farther in our course we shall describe archeological findings not only there, but in the Gobi Desert, Tarim Basin, etc. Lake Baikal is the deepest in the world (5300 ft.), containing one-fifth of the earth's fresh water, a lake formed by a huge earth crack 25 million years ago. According to UCLA Anthropologist R.D. McCracken grains were developed into food only 10,000 years ago, which Chinese records show were first used in the Gobi Desert and Tarim Basin, causing disaster in the formerly fertile Cradle of Civilization. In places, the desert of North Africa is even now advancing 30 miles per year; Formerly North Africa was once the breadbasket of the ancient Roman Empire. As man abandoned natural foods, not only the earth's face and form became more ugly, but likewise, the Races of man became uglier and degenerate.

MYSTICAL ANTHROPOLOGY
CONSTITUTED AS A FUNDAMENTAL
VITALOGICAL SCIENCE
by Dr. Johnny Lovewisdom

CHAPTER III   THE HEBRAIC ORIGIN OF THE OLD TESTAMENT BIBLE

"TO THE GOD OF RIGHTEOUSNESS, SHAMASH
At thy rising the gods of the land assemble, Thy terrible radiance overwhelms the land. From all lands together resounds, as many tongues; Thou dost know their designs, Thou dost behold their footsteps; Unto thee do all men look up together. Thou causest the evil-doer to tremble; Out of the depths Thou bringest those who perverted justice. Oh Shamash, by the just judgment which thou dost speak, Thy Name is Glorious; Thou standest beside the traveler whose way is toilsome; To the voyager who fears the flood thou givest courage. On paths that were never explored Thou guidest the searcher; Oh Shamash..."

    This poem, some 200 lines in length represents the sun's activity as a mysterious force by means of which all wickedness is frustrated, and all wrong-doers are brought to the book...There are innumerable analogies and homologies between the Old Testament, and the far older Babylonian writings, and these are most plausibly explained by the known fact that that Jews were exiles in Babylon, from 586 to 536 B.C., when they gave the final form to their "Mosaic" books. The stories of Creation, the Flood, the Confusion of Tongues, and the like, seem unmistakably derived from Babylonian sources. True the Jews amended all they borrowed, turning coarse myths into moral preachments and magical rites into pious symbols... The great Code of Hammurapi is the oldest collection of laws known to man... Tho many gods continue to be recognized, there was increasing reverence for only one, most notably the spirit of the sun, Shamash." The above poem, the lines that follow and the rising Shamash illustration I have taken from Lewis Browne's "The World's Great Scriptures", to show how modern spiritual concepts still remain the same as those of Sumerian origins of Babylonia dating at least to 3000 B.C., if not some eight to nine thousand years ago, which the Hebraic origins of Christian and Islamic Scriptures grasp only in spirit, beyond their chronological "beginning" of man's existence.

"This legend became known to the Jews from Chaldean sources, it was not known in Egypt, the country out of which they evidently came... Egyptian history had gone on 10,000 years before the time assigned for the birth of Jesus. It is known as an absolute fact that the land of Egypt was never visited by other than its annual beneficent overflow of the river Nile. The Egyptian bible, which is the most ancient of all holy books, knew nothing of the Deluge. The Phra, or Pharaoh, Khoufou Cheops was building his pyramid, according to Egyptian Chronicle, when the whole world was under the water of a universal deluge, according to Hebrew chronicle." Thus, a century ago, T. W. Doane substantiates the affirmations, quoted from numerous authorities on this subject, rather than depend on my own findings, or a charismatic "one and only God" that so many would rather see in the Bible as their Oracle for worship. It is so easily dismissed as a lack of Faith, as one idly lingers in ignorance, seeking to guard against any disturbance of one's complacent Salvation,- and avoiding attacks by "evil spirits", forever loyal to the Kingdom Gods and Babylonian path of religious life of one's paper oracle. However, how did those old Babylonian Scriptures describe the legends of Genesis, which have become the basic conception for present and even future thinking of many?

Before the reign of Sargon about 3,800 B.C., Babylonia had succeeded in forming itself into a national body, all speaking the same tongue, with the same customs and alphabet. Thus, the Hebrew Genesis had to start with the records made available to the Jews in their captivity, in divergent views always available, and speculation, like more than One and Only God, Bible and Faith divides people into quarreling groups, so people accept the most unreasonable and disgusting articles of religious faith, just to remain in the complacent calm of accepted beliefs. The Hebraic legendary author, Law-Giver, Moses, obviously describes Sargon, and the legends prevailing in this time. Here is his birth story:

"My mother of noble (poor?) family, conceived me and gave birth to me secretly. She put me in a basket of shurru reeds and shut up the mouth of it with bitumen; she cast me into the river, which did not overwhelm me. The waters carried me to Akki, the drawer of water. Akki reared me to boyhood." This is taken from an inscription written on a statue of the great king. This Babylonian origin of Moses is most logical, since the Biblic Genesis legends he describes are likewise also Babylonian legends, since in Egypt he had no written documentation of the Genesis legend of the Flood. What better source for data for this might we choose that of "The Historian's History of the World, Vol. I (Copyright 1904) quoted from Berosus: (Epoch of Xisuthrus: Over 64,000 B.C.)

"After the death of Ardates, his son Xisuthrus reigned 18 years. In his time happened a Great Deluge. The history is thus described: The Deity, Cronus, appeared to him in a vision, and warned him that upon the 15th day of the month Daesius (May and June) there would be a flood by which mankind would be destroyed. He therefore enjoined him to write A HISTORY OF THE BEGINNING, procedure and conclusion of all things; and to bury it in the city of the Sun at Sippara; and to build a vessel, and to take with him into it his friends and relations; and convey on board everything necessary to sustain life, together with all the animals, both birds and quadrupeds, and trust himself fearlessly to the deep. Having asked the Deity, whither he was to sail, he was answered "To the Gods": upon which he offered up a prayer for the good of mankind. He then obeyed the divine admonition:... After the flood had been on the earth, and was abated, Xisuthrus sent out birds from his vessel, which not finding any food, nor any place whereupon they might rest their feet, returned to him again. After an interval of some days he sent them forth again, and now they returned with their feet tinged with mud. He made a trial a third time...and the birds returned to him no more...He made an opening and found the vessel was stranded on the side of some mountain; he immediately quitted it with his wife, his daughter and the pilot. He paid adoration to the earth, constructed an altar, offered sacrifices to the Gods and with those who came out of the vessel with him, disappeared. They who remained within, finding their companions did not return, quitted the vessel with many lamentations an called continually on the name of Xisuthrus. HIM THEY SAW NO MORE; but they could distinguish his voice in the air, and could hear him admonish them to pay due regard to religion; and likewise informed them that it was on account of his piety that he was TRANSLATED TO LIVE WITH THE GODS; that his wife, daughter and pilot had obtained the same honor. They should return to Babylonia, and it was ordained, search for the writings at Sippara, which they should make known to mankind: moreover, that the place wherein they then were, was the land of Aramenia (Ararat in Hebrew)." (They found the writings, and Babylonia, once more, erecting temples, and from then on there are well known records.)

This we have quoted from accounts given by Berosus, who was born in Babylon in 330 B.C., thus being a contemporary of Alexander the Great, and according to Tatian, Berosus was the most learned of all Asiatic historians. Josephus quoted him to confirm his Jewish history, just as did Clement of Alexandria, and Eusebius used it to authenticate Christian Bible and Church origins. Berosus obtained his data from clay tablet libraries found in many large cities of Chaldea; what more, described by Pliny, the Athenians erected a gold-tongued statue in the Gymnasium, honoring his wonderful predictions as an Astrologer. The British Museum conserves clay tablets that give the same story told by Berosus altho not coinciding literally in parts of the accounts.

Now, we have arrived at a crucial point of confusion that illustrates the Sacred Mysteries and their interpretation, relating to the confusion of tongues and knowledge of Truth. The secular mundane Scientist seeking materialist facts, as well as the lazy brained sectarian religionist seeking to convert the whole world into their own convenient path of Faith in idolatry, with their paper God oracle or holy book interpretations, become willing multitudes that ruling political leaders manipulate and force into bloody wars seeking to monopolize the whole of mankind's body, mind and soul due to this lack of communication and communion due to this confusion that leads to chaos.

The Gnostic Hierophant, a Master of the Esoteric, is able to discern the Real Truth. The defenders of the literal interpretation of the Hebraic Biblic Flood, like Josephus and Eusebius, etc. took only what was needed for their purpose, ignoring the part I capitalized especially about Xisuthrus (or Chaldean "Noah's") statements describing this story as the History of Beginning, procedure and conclusion of all things, that They Saw Him no more, and rather than living after the boat trip again on earth soil, HE WAS TRANSLATED TO LIVE WITH THE GODS. Rather than authenticate the historicity of the Biblic Deluge. Like Josephus, they are ready to quote that, "Now all the Barbarian writers make mention of this Flood and the Ark; among whom is Berosus the Chaldean; for when he describes the circumstances of the Flood, he goes on thus:- "It is said there is still some part of this ship in Armenia, at the mountain of the Corydaeans; and some people carry off pieces of the bitumen, which they use chiefly as amulets for averting of mischief." Like our "Historians' History of the World", may we add,- "Prior to 1859, the people of Christendom rested secure in the supposition that the CHRONOLOGY OF MAN'S HISTORY was a fully known year of his creation. One had only to turn to the First Chapter of Genesis to find in the margin the date 4004 B.C. recorded with all confidence as the year of man's first appearance on earth." Chaldean records date from 500,000 B.C.

However, we must remember that Moses, supposed author of the Bible story of the Flood is of Egyptian origin, and Egyptian and Chaldean Mysteries, and hieroglyphic and cuneiform written accounts, in turn, all come from one Eastern Asiatic source. The keys to the Chaldean Mysteries, which Berosus coming from a priestly family and versed in Esoteric Sciences, quoted,- can easily be found in the Egyptian Bible, for which we refer to "THE BOOK OF THE DEAD" (English translation by Sir E. A. Wallis Budge. Chapter CVIII describes where Souls travel across the Celestial Seas on a vessel laden with all one's earth life food, tools and loved ones, including animals, to arrive at the home of the Gods in the Elysian Fields, "Sekhet-Aaru", home of the Blessed Dead.

"Now the Mountain of Bakhau (meaning Sunrise, located in Armenia, with Caspian seaport Bakhau east of it), whereupon this heaven supports itself, is situated in the Eastern part of Heaven, and it hath dimensions of 300 khet in length and a 150 khet in breadth. Sebek, the Lord of Bakhau to the East of the Mountain, and his temple is on earth there. There is a serpent on the brow of the Mt. and he measures 30 cubits in length; the first 8 cubits are covered with flints and shining metal plates. The Osiris Nu, triumphant knoweth the name of the SERPENT which dwelleth on that hill,- Dweller in Fire, is his name. Now after Ra hath stood still, he inclineth his eyes toward him, and the stoppage of the Boat of Ra taketh place, and a mighty sleep cometh upon him that is on the boat, and he gulpeth down seven cubits of the great waters. Thereby he maketh Suti (Set, personification of Darkness) to depart, having the harpoon of iron in him, and thereby he is caused to throw up everything which he hath eaten, and thereby Set is put into this place of restraint and then recite before him the enchantment saying, Get thee back to the sky, for that which is in my hand is ready." Thus, like in the "Healing God Spell of St. John", the evil satanic spirit is frightened away with sharp weapons (see Chapt. XXXII), and the doctrine of the Gnostics, one becomes "a Master of the Serpents of Ra", Ra being the Egyptian Sun God, and Serpents are the Forces of Darkness that the Sun overcomes. "I go round about Heaven, but thou art fettered by the fetters, which thing was ordained for thee formerly when Ra set in life in his horizon." Such are the funeral rituals of the deceased.

Reviewing these Egyptian rites, we add Chapter CLXXXVI: "Hathor, Lady of Amentet, mighty dweller in the funeral mountain, Lady Ta-ches-ert, daughter (or eye) of Ra, dweller before him, beautiful of face in the Boat of Millions of Years, the habitation (or seat) of peace, the Creator of Law (Law-giver) in the boat of the favored ones.... In this verse, there is evidence of where the titles of Moses, the Law-giver, and God's Chosen People, were copied from, while the "Ark of the Covenant", the "Ark of Noah" and the basket or Chest called "Ark of Moses", all represent the Sanctuary, Reliquary or symbolically the Spiritual Part of Man which is also the "Boat of Millions of Years" with which a man, together with his wife (two of every kind) travels to Eternity. Noah in Aramaic means Rest, Peace, Quiet or Calm which in the Ark (or Spiritual Part of Man) travels to Salvation, Immortality, thus is saved from the Race of Wicked Thoughts or Sin, to Rest in God or Spirit Forever. Noah was the son of Lamech (meaning strong young man) which with Noah (Rest) signify the Birth of a New Race, so that the allegorical Old Testament Scripture of Noah's Ark seeks to Cleanse the Consciousness, freedom from negative conditions and the Rainbow, Bow in the Clouds of Heaven or the ARC (Arch) is the sign of the Covenant between God and the Earth, of Salvation. "The Rainbow is the sign in the Heavens symbolizing the perfect blending of Race into obedience to one Harmonious Christ Principle" (Unity's Metaphysical Bible Dictionary).

The Ark of the Covenant is a Copy of the Ark of Isis, the Veiled Isis uses her wings as the veil, while her Ark is decorated with two Cherubs, the origin of Christian "Angels", and the pendant worn by the Hierophant or Chief-Priest of Initiation in the Mysteries, RA and Thmei, meaning the two Truths, are copied in the Jewish Priest's Urim and Thummin, of same significance, with 12 precious stones around this golden mirror represent the 12 Tribes of Israel, which were derived from the 12 sings of the Zodiac, like the 12 Apostles of Christ. The Khepera, Creator of Gods, is a form of the Rising Sun (Bakhau was the Mt. Ararat where the Sun rose), and Emblem is the Beetles worn on the Hierophant's chest, representing the Resurrection of the Dead unto Immortality, which correspond to the oracular images worn in Judean Priest's breastplate, used to mirror or divine the Truth. The origin of the Breastplate coming from the Disks worn by Egyptian figures representing the golden sun-God, Ra, or Ur in Aramaic, of which Urim is the plural, or Lights or Revelations, compared with Thummin, the whole or perfect Truth, is lost and unknown to modern priests in practices of rituals other than symbols.

"As a whole, THE BOOK OF THE DEAD, was regarded as the work of the God, Thoth, Scribe of the Gods, believed to be of Divine Origin," wrote Translator Sir E. A. W. Budge, and Thoth, the Atlantian is only describing the Mysteries from Atlantis. Thoth is described as Hermes of the old Greek myths, personified by Seth of the Hebraic Biblic legend. "Thoth with the Seven-Rayed Solar Discus on his head (like halo of Christian Saints) travels in the Solar Boat, 365 degrees, jumping out every 4th or Leap Year for one day". (Blavatsky) His Inner Eye, like those of all Atlantians is well developed, also symbolized by Enoch of Biblic Patriarchs according to the Masons, which gave the apocryphal Book of Enoch. The Ibis, sacred to Isis, is also sacred to Thoth, for the ibis kills land serpents, and eats crocodile eggs, saving the Nile from being infested by these animals, from which we get the Caduceus or

Staff of Hermes, brazen serpent of Moses, the Serpent of Wisdom symbol of Gnostics...But the mystical bird, ibis, if forgotten makes Wisdom worthless, so remember the bird, the Dove of Noah's Ark, Holy Spirit descending on Christ at the Jordan, the Swan of Eternity (Kala-Hansa) of Yoga, and allegorical theme of "Wise as Serpents, Harmless as Doves", required of Initiates of Immortal Godhood. Thoth, the Atlantian, brought us LOGOS WISDOM in the Egyptian and Chaldean Mysteries, which are the lost keys to Mysteries forgotten in modern religions, or of veiled meanings. "Among the Egyptian Gnostics, it was Thoth or Hermes who was the chief of the Seven, whose names are given by Origen as ADONAI, genius of the Sun; TAO of the Moon; ELOI, of Jupiter; SABAO, of Mars; ORAI, of Venus; ASTAPHAI, of Mercury; and ILDABAOTH (Jehova), of Saturn." (H.P.B.) These Lords of the Spheres and Zodiac, 7 rays of the Solar Disk, symbolize the Eastern Dhyani Chohans, Lords of Wisdom, who have taught Humanity all it knew. Each of the 7 Lords correspond to the 7 Ages of Kalahansa (or Time).

The preparing of Egyptian mummies was humorously relevant to my innocent attempts to eternalize the ever-changing body, in that in the process they removed the contents of the abdominal cavity, and then filled it with preserving essences of plants, and in some cases honey with its preserving essences was used for mummification. In my various schemes for escape from eating, hence working, and Mortality, after my 40 day fast at Lake Quilotoa (or Ark of 40 day Flood of Noah), a bit stunned by difficulty and lack of expectations, I remarked to students visiting from Quito somewhat as follows: Since during the fast the body continued to feed upon the putrid intestinal contents, in spite of continual high enemas which to the last days brought out fetid stinking wash water, gave toxic paralysis of strength of muscles, dizziness, a very coated tongue, dark cloudy urine, etc., a better plan would be to clean out the intestines with a fruit diet before a fast, and then fill the internal cavity with honey and then go on the Fast for Immortality! One would thus have enough fuel in food calories to last until one fairly forgot how and what it needed to eat for, and ipso facto it would come of itself, one became a Helioarian living from direct Sunlight. There seemed to be an ominous foreboding of events which led me not to try that. Later, more seriously I contemplated going to live in the snows of Mt. Chimborazo, and since all hunger leaves one at those snowy super-altitudes, and since if one's attempt fails, one cold freeze to death,- like other fatal victims of freezing, one passes to Eternity with a broad Buddha like smile or grin, -success would be guaranteed! There seemed to be a great resemblance of this poor hermits escape to Immortality on the Boat or Ark of Millions of Years, to the lavish designs of Monarchs of old. The welfare of the dead was more important to the Egyptians, than that of the living, due to the fear of the hereafter. Life was often sacrificed to obtain petrified immortality in sterile mortuary palaces. Later, we shall study how in Krete and among Greeks the cult of the living was restored, replacing the epochal cult of the dead.

Let us return again to the subject of this Chapter, the Hebraic Origin of the Christian Bible. In spite of knowing that the Hebrew Doctrine, the Ten Commandments (Chapt. CXXV, Book of the Dead) and the major part of Jewish Religion was copied from Egyptians and Chaldeans, who in turn received it from a sunken continent we are collecting a pre-history about, we do have an excellent source even among so-called Essene students. It would be ambiguous to ask Paradisians to believe:- a Sacred History written and re-written by the Hebraic Temple Slaughter-house cult and Holy Roman Empire promoters, who actually believed like the Babylonians, that the slaughter of animals and men, or bloodshed, can save a soul from sins, as claimed by Jewish and Christian teachers. Flavius Josephus is a historian who wrote about the Jews that there has been no doubt as to if truly he did exist, and he was known to exist contemporary to the time of the New Testament Apostles, altho he had no knowledge of a N.T. Scripture in his time,- 37 to 95 A.D. All the Christians before Eusebius (Church historian of the 4th Century), held the view, like Origen, Clement of Alexandria, Justin Martyr, etc. that Josephus, who had mentioned John the Baptist, did not acknowledge Christ. When the Christian Religion was legally organized in the 4th Century, Eusebius was the first to quote a new spurious passage summarizing Christian doctrine and existence. But note well, he also wrote: "How far it may be proper to use FALSEHOOD as a medium for the benefit of those who require to be deceived,...I have suppressed all that could tend to disgrace our religion", and thus held it is lawful to cheat and lie for the cause of Christ! This reveals why book burning by the Roman Church was able to destroy almost all evidence against Rome's dogma.

Otherwise, "THE WORKS OF FLAVIUS JOSEPHUS" (see translation by Wm. Whistan) are a good source to find historic confirmation of the principles we teach in the Vitalogical Sciences. Like Herod Antipas, in Mark 6:14, states of the appearance of the Savior ("Jesus" is word for the Savior), "John the Baptist is risen again from the dead, and therefore mighty works show forth in him." Even the Gospel of this Savior, John reborn of Spirit and Living Water, emphasizes so no one will be misled, "Verily I state the Truth, there hath not risen among them that are born of women, a GREATER THAN JOHN THE BAPTIST" (Matt. 6:11, Lk. 7:28). Thru-out the Gospels, clues are given that they are only to be used as an allegorical representation of Truth to the Spiritually Initiated.

It would require duplication of Josephus's Work beyond our capacity herein, so I will quote only the passages of vital interest to us here in our book: "Moses says further, that God planted a Paradise in the EAST, flourishing with all sorts of trees; and that among them was the tree of life, and another of knowledge, whereby was to be known what was good and evil; and that when he brought Adam and his wife into this garden, he commanded them to take care of the plants. Now the garden was watered by one river, which ran round about the whole earth, and was parted into four parts ...God commanded that Adam and his wife should eat of all the rest of the plants, but to abstain from the tree of knowledge, and foretold to them that if they touched it, it would prove their destruction. But while all the living creatures had one language at that time, the serpent, which had lived together with Adam and his wife, showed an envious disposition, at his supposal of their living happily, and in obedience to the commands of God; and imagining that when they disobeyed them, they would fall into calamity, he persuaded the woman, out of malicious intention, to taste of the tree of knowledge, telling them that in that tree was the knowledge of good and evil; which knowledge when they should obtain, they would lead a happy life, nay, a life not inferior to that of a God: by which means he overcame the woman and persuaded her to despise the word of God. Now, when she had tasted of that tree, and was pleased with its fruit, she persuaded Adam to use of it also. Upon this they perceived that they were become naked to one another; and being ashamed thus to appear abroad, they invented somewhat to cover them, "for the tree sharpened their understanding, and they covered themselves with fig-leaves, and tying these before them, out of modesty, they thought they were happier than before, as they had discovered what they were in want of. But when God came into the garden, Adam, who was want before to converse with him, being conscious of his wicked behavior, went out of the way.

"This behavior surprised God: and he asked what was the cause of this procedure: and why he, that before delighted in that conversation, did now fly from it, and avoid it. When he made no reply, as conscious to himself that he had transgressed the command of God: God said, "I had before determined about you both, how you might lead a happy life, without any affliction, and care and vexation of soul: and that all things which might contribute to your enjoyment and pleasure should grow up by my Providence, of their own accord without your own labor and pains-taking: which state of labor and pains-taking would soon bring on old age: and death would not be at any remote distance: But now thou hast abused this my good-will, and hast disobeyed my commands; for thy silence is not a sign of virtue, but of thy evil conscience." However, Adam excused his sin, and entreated God not to be angry at him, and laid the blame of what was done on his wife; and said that he was deceived by her, and thence became an offender; while she again accused the serpent. But God allotted him punishment, because he weakly submitted to the counsel of his wife; and said the ground should henceforth yield fruit of its own accord, but that when it should be harassed by their labor, it should bring forth some of its fruits, and refuse to bring forth others. He also made Eve liable to the inconveniency of breeding, and the sharp pains of bringing forth children, and this because she persuaded Adam with the same arguments wherewith the serpent had persuaded her, and thereby had brought him into a calamitous condition. He also deprived the serpent of speech, out of indignation at his malicious disposition toward Adam. Beside this, he inserted poison under his tongue, and made him an enemy to man; and suggested to them that they should direct their strokes against his head, that being the place wherewith lay his mischievous designs toward men, and it being easiest to take vengeance on him that way: and when he deprived him of the use of his feet, he made him go rolling along, and dragging himself upon the ground. And when God had appointed these penalties for them, he removed Adam and Eve out of the garden into another place." (Let us pause here, to study the meaning of the citation.)

First, let me cite two passages that preceded these scriptures: "God took dust from the ground, and formed man, and inserted in him a spirit and a soul." The translator thus added a footnote which stated, "Josephus supposed man to be compounded of spirit, soul and body, with St. Paul, I-Thess. V:23, and the rest of the ancients: he elsewhere says also that the blood of animals was forbidden to be eaten, as having in it soul and spirit." All of this allows for the doctrines we hold as to abstinence from bloodshed, which deprives humans and animals from their integration of body, soul and spirit, and eating of seminal substances such as nuts, grains, legumes and other seeds, which obviously furnish humans with reproductive substance, and overcame Eve and then Adam with overwhelming reproductive passion, losing their Paradisian Providence.

Also, there is the passage stating that God put Adam to sleep, and "took away one of his ribs, and out of it formed the woman; whereupon Adam knew her when she was brought to him, and acknowledged that she was made out of himself." This shows there can be children without the sexual means of giving birth, of an androgynous nature. Moreover, the whole exposition given by Josephus is typically Gnostic, and since the First Christians were known to have been disciples of St. John, and the Christians of St. John were called Gnostics, Nass-Aryans, Mandaens, etc. which means that Josephus' writings which we are quoting instigated all the alternate Gnostic doctrines that followed immediately following him in the 2nd Century. Josephus' generosity with the Serpent in making him a partner with the first couple, as well as a teacher who was able to guide them into knowing Godhood, or Gnosis, rather than live in blind Faith, and ignorance of good and evil, was in time to be fully exploited by Gnostic Doctors of multiform doctrines. The Serpent had a soul and spirit like man, in body walking and talking like him, so not only could he equate to humans, but moreover he had evidently been freed from ignorance, having become a knower of Gnosis, Godhood. The orthodox or dogmatic opinion of Jews and Christians was that one should live by Faith without seeking reason and logic (Logos) for Truth, which made Hermetic teachings (of Thoth, the Atlantian), as well as St. John, characteristically "Gnostic". As Mdme. Blavatsky continually indicated, "The Gnostics, or early Christians, were but the followers of the old Essenes, under a new name". Buddhist missionaries to Egypt and Palestine spoke of their Sages as "Nagas", Kingly Serpents, so likewise the Gnostics had their Ophites, Nassenes, Nass-Aryans and other Serpent cults, beside Kundalini Masters of Yoga, Ancient Hyperborean Sciences. Wisely enough Josephus stated above that Paradise was located in the East, beside a place where fig leaves had to be replaced by animal skin and furs. As Epiphanius revealed about the Ophite Gnostics, they honored the Serpent because he taught the primeval men the Mysteries. Thus the SEVEN HEADED SERPENT, Seven Logoi, becomes the Key to Genesis when like the symbolical Kabala we identify the Serpent with Lord God, or "Demon est Deus inversus" (Devil is inverted or reverse of God). "Object of horror and of adoration, men have for the Serpent implacable hatred, or prostrate themselves before his genius. Lie calls it, Prudence calms it, Envy carries it in its heart, and Eloquence on its caduceus. In hell it arms the whip of Furies, in Heaven, Eternity makes of it its symbol. (De Chateaubriand). The Seraphim are the fiery Serpents of Heaven who guard Mt. Meru, the Heavenly Mt., serpents that swallow their tails to become Winged Globes, Fiery Wheels or now "Flying Saucers". The Winged Serpent, Seraphim, of seven heads also representing the Elohim or Dhyana Chohans. "Then the Lord God said, Behold, man has become like one of us, to know good and evil: and now, lest he put forth his hand, and take also of the tree of life, and eat, and live forever," he was cast out of Eden. (Gen.3:22) Josephus, John's scribe, continues his exegesis:

"Adam and Eve had two sons: the elder of them was named Cain, this name signifying Possession. The younger was Abel, which signifies Sorrow. They also had daughters. Now, the two brethren were pleased by different courses of life; For Abel, the younger was the lover of righteousness, and believing that God was present at all his actions, he excelled in virtue; and his employment was that of a shepherd. But Cain was not only very wicked in other respects, but was wholly intent upon getting; he first contrived to plough the ground. He slew his brother on the occasion following: They resolved to sacrifice to God. Now Cain brought the fruits of the earth, and of his husbandry; but Abel brought milk, and the first fruits of his flock; but God was more delighted with the latter oblation, when HE WAS HONORED WITH WHAT GREW NATURALLY OF ITS OWN ACCORD, than he was with what was the invention of a covetous man, and gotten by forcing the ground; then it was that Cain was very angry that Abel was preferred by God before him; and he slew his brother, and hid his dead body, thinking to escape discovery."

We pause here to condense a bit, to explain that God said "I wonder at thee, that thou knowest not what is become of a man whom thou thyself hast destroyed." So Cain and his wife are cast out of that land, just as Anthropology also has traced civilization ever westward from its origin in North-East Asia, and while you read it, see how it's the identical story that in Chapter II we tell of the Buddhist Genesis.

And when Cain had traveled over many countries, he with his wife built a city named Nod, which is a place so called, where he settled and he had children. However he did not accept punishment in order to do amendment, but to increase his wickedness; for he only aimed to procure everything that was for his own bodily pleasure, tho it obliged him to be injurious to his neighbors. He augmented his household substance with much wealth, by rapine and violence; he excited his acquaintance to procure pleasures and spoils by robbery, and became a great leader into wicked courses. He also introduced a change in that way of simplicity wherein men lived before; and was the author of measures and weights. And whereas they lived innocently and generously while they knew nothing of such arts, he changed the world into cunning craftiness. He was the first of all to set boundaries about lands; he built a city and fortified it with walls, and he compelled his family to come together to it; and called that city Enoch, after the name of his eldest son, Enoch." (We again pause to avoid repetition of the Bible's Genesis).

As Cain's children got wicked also they invented musical instruments and weapons working with brass, and spread their wicked way of life. Adam and Eve had many other children, 33 sons and 23 daughters, an old tradition relates,- but they had one son who became notably virtuous. This was Seth, and they continued to inhabit the same country as Adam, "without dissensions, and in a happy condition, without misfortunes, falling upon them until they died. They also were the inventors of that peculiar sort of wisdom which is concerned with the heavenly bodies and their order. That their inventions might not be lost before they were sufficiently known, upon Adam's prediction that the world was to be destroyed at one time by the force of fire, and at another time by the violence and quantity of water, they made two pillars; one of brick and the other of stone: they inscribed their discoveries on them both...and they remain in the land of Siriad to this day."

Now, this is exceedingly interesting, since we have the FIRST BOOK OF ADAM AND EVE with 79 Chapters and the SECOND BOOK OF ADAM AND EVE with 22 Chapters (The Forgotten Books of Eden) which describe just as Josephus does about Adam and Eve, and Seth, and their children who lived on Mt. Hermon on the southern border of Syria as described above. This means Josephus of the first century of our Era, was either the author of these ideas that were preached in the Apocalypse and the Adam and Eve Book, or more likely, he is briefly adding some of these ideas derived from Gnostic texts that existed before his time, rather than there being any truth in the Christian Church accusation that the Gnostics had invented them in the 2nd and 3rd Centuries. As to the confirmation of our doctrine that grain and seed eating caused man's degeneration, in the Book of Adam and Eve (R. H. Platt version), we read:

"Seth and his children dwelt on the mountain below the Garden; they sowed not, neither did they reap; they wrought no food for the body, not even wheat; but only offerings. They ate of the fruit and of trees well flavored that grew on the mountain where they dwelt. Then Seth often fasted every 40 days, as did his eldest children. For his family smelled the smell of the trees of the

Garden, when the wind blew that way. They were happy, innocent without sudden fear, there was no jealousy, no evil action, no hatred among them. There was no animal passion..." They lived in perfection, as long as they mingled not with the children of Cain, altho in time, even they were drawn away by sweet music, dancing and drinking to mix with the children of Cain on the plain west of the Holy Mountain. "And the Word of God said unto Adam, As to the fruit of the Tree of Life, for which thou askest, I will not give it thee now, but when 5500 years are fulfilled. Then will I give thee of the fruit of the Tree of Life, and thou shalt eat, and live forever, thou and Eve, and thy righteous seed."

Returning to Josephus, we add one more passage that seemed very significant: "Now when Noah had lived 350 years after the Flood, and all the time happily, having lived 950 years, he died: But let no one, upon comparing the lives of the ancients with our lives, and with the few years which we now live, think that what we have said of them is false; or make the shortness of our lives at present an argument that neither did they attain to so long a duration of life; for those ancients were beloved of God, and made by God himself; and because their food was then fitter for the prolongation of life, might well live so great a number of years; and beside God afforded them a longer time of life on account of their virtue, and the good use they made of it in astronomical and geometrical discoveries, which would not have afforded the time of foretelling the periods of the stars, unless they had lived 600 years; for the Great Year is completed in that interval. Now I have for witnesses to what I have said all those that have written Antiquities, both among the Greeks and barbarians; for even Manetho, who wrote the Egyptian History, and Berosus, who collected the Chaldean Monuments and Mochus, and Hestiaeus, and beside these, Hieronymus the Egyptian, and those who composed the Phoenician History, agree with what I say: Hesiod also, Hecataeus, Hellanicus and Acusilans; beside these, Ephorus and Nicolaus RELATE THAT THE ANCIENTS LIVED A THOUSAND YEARS: but as to these matters, let everyone look upon them as he thinks fit."

Altho Jews were thought to originate in Chaldea, one of our collaborators sent us the following data: The important book

"THE HEBREW TONGUE RESTORED" by Olive d' Fabre has been reprinted by Samuel Weisser of New York from the 1889 original. The writer proves that the Hebrew language was not a spoken language, but a secret Egyptian code, that was written but not spoken, in the same way symbols are used. This language code was created by the high priests of Egypt to convey their doctrine and truths. In other words, the language that is being passed off as Hebrew and that its history is very ancient, is a FRAUD. The writer confines himself solely to the ethnology and construction of the Hebrew tongue and finds that its origin is found in Egypt, and not anywhere else. The Egyptians were not related ethnically to the African savages. They were Aryans, Newcomers from Mesopotamia, Asia and India as were the Hyksos kings. Immediately beyond the borders of Egypt began Nubia, the territory of the native African tribe called the Blemmyae, a tough people and province. Ammianus Marcellinus, the Roman historian (360 A.D.) says the robber-brigands who periodically had over-run Arabia, thru the Negev of today, from "the borders of the Blemmyae to Syria to Gaza and about the Dead Sea,- were spawned in Ethiopia, meaning the Sudan, Nubia, the Blemmyae. In this region was and is the island community called Elephantine, some 700 miles up the Nile, where the mixed multitudes of the ancient Iebrons (Hebrons) were "spawned" and where Moses found his "Ethiopian wife". This, then is where the ancient Hebrons found their wives. Jacob who went into Egypt with no more than 70 souls, soon increased into a tribe of 600,000. The Bantu Pygmy race of Central Africa had colonized Blemmyaes, and they bred the vast increase mixing with the Hebrons at Elephantine, where a trade center was established between the Egyptian Delta and the Sudan. The Egyptians refused to marry with the original Hebrons because they had become so unclean with disease, which caused them to develop so many "unclean" sexual restrictions and to this day they are distinguished as those who have to circumcise, because of the racial mixing in Africa. The Old Testament history in Egypt began by reversal of facts, as commonly known, to justify all the crimes they depended upon for survival. As Kephart shows, in Babylonia those under A-Braham (against Brahm) gathered into a tribe of Sa-Gas, robber-brigade in meaning, which were expelled from Elam, then from Ur, then from Canaan, then Egypt, "King David owed his success to the employment of a troop of freebooters", yet all their actions were recorded as victories, history contradicting often, as accomplished with Yahweh's favor to them. The Bantu Pygmies called themselves the Kozan peoples, meaning the Koza or Select, which the Bible calls Chosen People, with Kosher foods and rituals. The Kozan people attached the Egyptian Coded Hieroglyphics to their own speech sounds. Ethiopia was founded by Solomon's first son, born by the Queen of Sheba.

One might find justification as to "unclean" seed in Numbers 12:1 "Miriam and Aaron spoke against Moses because of the Ethiopian women whom he had married; for he had married an Ethiopian woman...Behold Miriam became leprous, white as snow." Josephus describes how when Moses was made a general for the Egyptians, he made "a great slaughter of the Ethiopians", but spent 9 years without being able to conquer them, so "Tharbis was the daughter of the king of Ethiopians...admired the subtlety of his undertaking...she fell deeply in love with him." Thus, they made a contract, the city would deliver itself up if Moses kept his oath to her. "No sooner was the agreement made, but it took effect immediately; and when Moses had cut off the Ethiopians, he gave thanks to God, and CONSUMMATED HIS MARRIAGE, and led the Egyptians back to their own land." When said of marriage, Webster defines consummate as "to complete with intercourse." Josephus tells how Moses uses the Egyptian belief that the ibis is sacred because it eats serpents, to win battles, and when he also imitated the Egyptian magicians by turning rods into serpents, claiming this came by Divine Power, the Egyptian king and Moses accused each other of arts and tricks. "The king was angry with him and called him an ill man, who had formerly run away from his Egyptian slavery, and came now back with deceitful tricks, and wonder in magical arts to astonish him." There is very little in the so-called histories of Moses, Jacob, and other great Jews that is not exploiting, trickery, cunning and violation of the commandments they boast to give to mankind. Another variation of this is given in the Book of Jaspar 72:37, saying when King Kikianus died, "All the peoples and nobles swore unto Moses to give him for a wife, Adoniah, the Queen, the Cushite wife of Kikianus, and they made Moses king over them on that day, ... Moses reigned 40 years over the Cush." After this Moses is said to have married

Zipporah, a Midianite, Midianite means Strife, petty quarrels, jealousy and uncharitable thoughts. After the two Ethiopian wives, and troubles with Zipporah, Miriam, we can add little!

In turn, the White Israelite movement claims, "The Jews became Jews about the 7th Century A.D. when the King of the Mongoloid Khazar Empire (near the Black Sea) accepted Judaism and made it the State Religion and these Khazar Jews were never in the Holy Land. This they hitch to theories that they were a serpent race, due to Eve's seduction by the Serpent that walked and talked like a man, which gave the birth of the descendents of Cain, while Abel was from the seed of Adam! We have no quarrel or "Midianite" certainty about any of this, and at best have only found groups that were seeking to support their own racial background.

Much more important is the Spiritual Heritage that might be derived from scriptural concepts. Mdme. Blavatsky affirms, "In spite of all controversies and researches, History and Science remain as much as ever in the dark as to the origin of the Jews. Yet whatever they may have been, they became a hybrid people, not long after Moses, as the Bible shows them freely intermarrying not alone with the Canaanites, but with every other nation or race they came in contact with." She also held that "the story told of Moses by Ezra had been learned by him while at Babylon, and that he applied the allegory told of Sargon to the Jewish Lawgiver. In short, that Exodus was never written by Moses, but re-fabricated from old materials by Ezra." J.C. Bonner (New Light from an Old Lamp), adds, "At first only the Torah which we have been erroneously taught to call the "Five Books of Moses", was translated into Greek from a Samaritan text. Moses of course was not the author of the Torah as the text itself discloses for how could a dead man write his own biography? This translation was made in Alexandria, Egypt, in 285 B.C." Bonner further establishes that the Old and New Testaments are not two different Testaments, but that the N.T. is the "Testament Anew" or Restored. In the writings of Dr. Anna Kingsford and Edward Maitland, "The Perfect Way", they confirmed findings of Blavatsky and others, quoting Mosheim: "All the Fathers of the second century attributed a hidden and mysterious sense to the Scriptures." Papias, Justin

Martyr, Irenaeus, Clemens Alexandrinus, Gregory of Nazianzen, Gregory of Nyssa and Ambrose, held that the Mosaic account of Creation and Fall were a series of allegories. History tells us, moreover that the Book of Genesis as it now stands, is the work, not even of Moses, but Ezra or Esdras, who lived at the time of the captivity,- 500 to 600 B.C. and that he recovered it and other writings by the process already described as intuitional Memory. "My heart", he says, "uttered understanding, and wisdom grew in my breast; for the Spirit strengthened my memory." If then by such means he recovered what Moses had previously delivered orally to Israel, it obviously is that Esdras must have been initiated into the ancient tradition in a former state of existence; since no MEMORY could have enabled to recover that which had never been known, and which,- when the Divine commission to rewrite it was given him,- was so wholly lost that "no man knew any of the things that had been done in the world since the beginning." As the Talmud says, "Ezra could not have received the word, if Moses had not first declared it." Neither must it be supposed that we have the Books of Moses as recovered by Esdras. "The system of interpolations and alternations already referred to as largely applied to the Bible, especially affected the Pentateuch. And foremost among those who thus perverted it were the Pharisees, denounced in the N.T., who greatly modified the text, introducing their own ritual into the law, incorporating with it their commentaries, and suppressing portions which condemned their doctrine and practice. According to Spinoza, "there was the time before the Maccabees, no canon of the holy writ extant; the books we now have were selected from among many others by and on the authority of the Pharisees of the second Temple, who also instituted the formulae for the prayers used in the synagogue." "The Perfect Way" goes on to say, "The great parabolic Myths which lie embedded in the Hebrew Scriptures (are) like so many gems encased in clay. And gems these are, which, from prehistoric times, have been the property of the Initiates of all religions, and especially the Hindu and the Egyptian, from which last indeed Moses originally drew them, as it is occultly intimated when it says, "And the children of Israel BORROWED OF THE EGYPTIAN JEWELS of silver and jewels of gold, and they spoiled the Egyptians." With regard to this particular Myth of the Fall, the walls of ancient Thebes,

ELEPHANTINE, Edfou, Karnak bear evidence that long before Moses taught, and certainly ages before Esdras wrote, its acts and symbols were embodied in the religious ceremonials of the people, of whom, according to Manetho, Moses was himself a priest. And "the whole history of the fall of man is," as says Sharp in a work on Egypt, "of Egyptian origin."

We have specially quoted form Dr. Anna Kingsford and Mr. Edward Maitland because in 1881 they were foreshadowing the Paradisian teaching: "That this is literal truth, all the poets, all the seers, all the regenerate testify, bearing witness also that Paradise can never be regained, Regeneration never completed, man never fully redeemed, until the body is brought under the Law of Eden, and has cleansed itself thoroughly from the stain of blood. None will ever know the Joys of Paradise who cannot live like Paradise-man (PARADISIANS!); none will ever help to restore the Golden Age to the world who does not restore it in himself. No man, being a shedder of blood, or an eater of flesh, ever touched the Central Secret of things, or laid hold of the Tree of Life. ...Lips polluted with blood may not pronounce the Divine Name...For if he will but live the life of Eden, he shall find all its Joys and mysteries within his grasp. "He who will do the will of God, shall know the doctrine. "Until 'father and mother' are forsaken, that is, until the disciple is resolved to let no earthly affections or desires withhold him from the Perfect Way,- Christ will not be found, nor Paradise regained." In the Preface to the Fifth Edition, S. Hopgood Hart, wrote:

"Precisely at the moment when Anna Kingsford and Edward Maitland were entering upon the collaboration which had for its aim the restoration, interpretation and vindication of the great mystical system of the West which underlay all its ancient religions and Sacred Scriptures, Col. Olcott and Mdme. Blavatsky,- founders of the Theosophical Society,- were preparing on the other side of the Atlantic, to do the same thing for the corresponding system of the East; and it is remarkable that the whole movement comprising these two events, and its rise precisely at the time for which it had been announced in numerous predictions from before the Christian Era to the later Middle Age." The Perfect Way is a Divine Revelation which has restored to the world "that famous system of cosmogony, which is known to Initiates as the

Hermetic Gnosis, has from the remotest antiquity been venerated as the one true Divine Revelation concerning the nature of man and the universe..."

MYSTICAL ANTHROPOLOGY CONSTITUED AS A FUNDAMENTAL VITALOGICAL SCIENCE by Dr. Johnny Lovewisdom

CHAPTER IV: DIETETIC VIOLATIONS THAT GAVE RACIAL TRAITS TO PREHISTORIC PALMIRS, TARIM BASIN, SUMERIA CHALDEA FROM ATLANTIS

ANTHROPOLOGY is the Science of Man, coming from "Anthropo", the Greek word for man, and "Logy" denoting Science. The Study of Cultural Anthropology includes ARCHEOLOGY,- the Study of extinct cultures,- and ETHNOLOGY,- the Study of living cultures and linguistics. In this Chapter we shall deal with the latter, that is Ethnic Origins and Etymology, or the origins of race and language.

Humans are all of the same refined Divine racial origin. It is not logical, nor psychologically sound, to believe that something such as very complex things can be derived from nothing, or that the effects have no cause, the lesser part is the same as the whole. It eliminates the bulk of man's problems when we realize that the Cause is Cosmic, and the effects are only local of this passing moment. Different racial divisions came about by grosser, coarser or degenerate racial characteristics brought about by man's unnatural and immoral living habits adopted by customs, traditions and environment which grew into hereditary traits in family, tribe and eventually nation. These differences created belligerent animosity in these clans, tribes, and nations, each one seeking to defend their favorite and convenient principles in ideals of living and claims to Divine origin, and contrasting this to the evils and criminal offences against Life and Divinity that all others outside their clannish and national ties practiced, justifying their destruction to prevent further human decadence among the human species. As Dr. Kephart summed it up in "Races of Mankind", "It is believed that differences in altitude, climate, diet, and environmental conditions account not only for differences of complexion and figure, but also for the character of the hair, the width of the nostrils, the cephalic index, and other physical variations in man."

As I have long maintained, words were invented to conceal the

Truth. To prevent the mingling or inter-marriage of races, each group adopted their own meaning to words, giving distinct or opposing meanings and sarcastic usage, and then the mockery of accents and dialects led soon to creating new words from the original Word or Speech of our One Origin. Indoctrination of differing cultural and spiritual principles, by clannish, tribal or political secrecy in linguistic barriers, ritual symbols, etc. divided the earth's inhabitants until civilization has become one chaotic force seeking self destruction. The Modern Materialistic Cult of Science, Politics and Business now continues in this self appointed gospel to extend their dominion over all else, blindly disregarding the ecological destruction of all mankind that artificiality in man's living habits involves, visible and evident to those involved. Each one defends their own convenient principles and condemn all else.

The Lovewisdom Message in the Vitalogical Sciences seeks to incorporate all the spiritual and natural living principles in a sublime and strict discipline in practice, harmonious to our singular racial and cultural origins, for a Paradisian racial and new age rebirth of mankind. As we have pointed out, the oldest cultural traditions and archeological findings indicate that our ethnic origin, or "Cradle of Mankind", was in northeast Central Asia of what has been referred to as a Hyperborean Culture in a Paradisian Environment which enabled mankind to subsist thru a frugivorous diet, livelihood and spiritual orientation.

However, what contrasts all other teachings, is our revelation that both the eating of seeds, principally grains, and the slaughter of animals for foods, rather than deliver mankind from sins (as Babylonian, Egyptian, Hebrew, Christian and other sacrificial doctrines hold in scriptural principles), is the very cause of man's degeneration into racial divisions, war, passion and suffering. For the verification of our teaching, I would like to present some scientific findings that the "SCIENCE AND PRACTICE OF IRIDOLOGY" brings out as the work of our friend, Dr. Bernard Jensen, renown authority on Chiropractic, Nature Cure and Drugless Healing. Let me quote random passages in his book:

"A toxic condition has much to do with the color of the iris just as toxins in the body will make the complexion darker. It is an interesting experience to see the eyes that have been classified brown become lighter, and the blue fibers become more predominant through your treatments... When a baby is first born his eyes are lightest in color. As he goes thru life he acquires toxins thru poor living habits and these result in a settlement of different fluids which cause changes in the color of the eyes...The eyes of the northern races are more or less of the true blue color and as we go south we come to brown-eyed people... In general, the degree of blond and brown pigmentation of the hair determines the degree of blue and brown pigment in the iris...Those who take care of themselves, whether brown-eyed or blue-eyed can be long-lived. In my practice, however, I find those who have lived the longest life have blue eyes...If you could have the opportunity of looking at the iris of a person immediately after he has been in an accident or has suffered from some great catastrophe, you would see the accident almost simultaneously registered in the iris...The chameleon can ably camouflage itself thru color changes that blend with the environment. An interesting fact, however, is that only one-half the body of this animal will change color if the sight of one eye is kept from observing its environment. Now we know the environment had no part in the physical change of color in that chameleon. It was produced by a vibratory change thru the nervous system. When we deal with the vibration of different diseases and the discoloration of the iris from drug deposits and bad living habits, we are dealing with electrons and ions... When the body must continually struggle against foreign substances, it becomes enervated and disease is bound to result. The activity is always shown in the iris... Diet has a definite vibratory effect on the body...Fruits have a higher vibratory action than vegetables. They stir up the acids in the body and bring on an acute form of elimination, while vegetables seem to carry off toxic material and are very soothing especially to the nervous system. Disease comes into existence thru abnormal vibrations...A person with a loss of substance in his body has little chance of overcoming sickness. Unless he abuses his body, the person with exceptionally good density has the opportunity of living a long life; he has the capabilities of overcoming disease. Density can be regarded as the measure of a body's vitality. When there is defective tissue density, the fibers in the iris are unevenly developed and arranged.. Taints are reflected in the iris not only from our grandparents, but even from our great, great, great grandparents."

Needless to say, we have quoted only parts we thought important to prove our thesis about racial origins, and not that Dr. Jensen teaches this in any way. Readers desiring to obtain the quoted book on Iridology may write Dr. Bernard Jensen, Route 6, Box 811, Escondido, Calif., 92025, U.S.A. If we take note of the racial and linguistic implications in the cognomens of the pioneers in Iridology, Ignatz von Peczely, Nils Liljequist, Johannes Thiel, Henry Lindlahr, etc. beside Dr. Jensen, one immediately notes the Northern or Nordic blue-eyed blond affinity which they worked with, and found such remarkable evidence in the iris therein. I did not start studying Iridology until I was in the high Andes using Spanish textbooks, without a full grasp of the Science at first. But as years passed and one observes the eyes of hundreds of people, the eyes soon become "the window of the soul", and especially the past physical condition of those we meet. In fact when I came to this Valley of Vilcabamba, I found so many blond or red haired, blue or green eyed inhabitants, that along with the great many centenarians, I soon concluded this was in ideal Health center, or Valley of Longevity.

However, I soon witnessed that in only a matter of 5 to 10 years after meeting distinctly blue-eyed people, their eyes had taken on muddy green to brown iris color, reflecting their diet of rice, bread and other processed foods that were brought in with better roads, beside the yucca-root, corn, animal flesh and fruits they had when the centenarians were but "babes in the woods" here. I have known many Health or Vegetarian minded parents who themselves had dark features, but as long as they raised their children on a diet of uncooked "raw" foods their children remained blond and blue eyed. Even when parents with black hair and complexions used milk, vegetables and fruit in abundance without avoiding some processed food and flesh, their children were blond and blue eyed until adolescence and then took on darker colors.

The most remarkable case came with my companion Ruth Marie, who arrived here in 1974 with blond hair but hazel eyes, as also described on her passport. She lived strictly on fruits and salad vegetables, usually dressing salads with clabber or avocados for over 4 years now. Her eyes have changed from hazel to clear blue, and she healed her former menstrual flow, beside developed a remarkably strong body.

It has been convenient in the past to pass these observations off as a matter of race, lack of pigment factors, etc., but if one will make an honest effort to study irises the facts become evident: the irises turn darker, the skin becomes darker and irregular with blemishes and other racial features by an increase of toxin-laden tissue in the body, coming from indulgence in cooked animal flesh, grain and other seed products of all kinds, drugs, and other body poisons. The only thing in common to racial color and features is generations of abuse so the children are born with toxin-engendered-pigmentation from their parents and immediate forefathers, and bred into habits of living, customs and traditions that continues building up darkness of soul and body, just as living in ignorance is characterized as "darkness". In my 1962 publication, "ORDER OF PARADISE", I observed, "Dead foods gave the black, death color to shades of race. African savages clubbed the animal or human until black and blue to kill it before cooking it, obtaining the darkest substance, like the black, dead flesh of gangrene. Asian and Andean Indians, starchy grain-eaters, took on a toasted brown color of fired grains and earth, the inorganic colors of petrification. The yellow urine color reflects the Chinese method of using excrement as farm fertilizer (beside use of legumes, grains, seafoods) due to the high uric acid content, also witnessed in the olive color of herbivorous animal flesh eaters. Yogis living on fruit become fair complexioned if dark. Negros living on uncooked green peas doing harvest work become ashy white. Cooked dead food does not polarize light,- a beautiful display of color anyone can watch training a microscope on living cells scintillating with sunlight. The minerals in live food act as magnets in holding the sun's energy, and fill our body with Solar or Sun Light which is Sun Light Power, True-born Sons of the Sun.... The civilized white people are fair and tall from milk foods, a biological triple bleach in the case of clabber milk (Chlorophyll is bleached in hay fed cows, and bacterial plants lack iron, making it imperative to use clabber with greens), just as headed and root vegetables lack the sunlight and iron factor. Statistics show a proportional increase of darkness of skin in grain and pulse consuming people which whitens when there is greater use of milk products.

However, altho I had to admit the above facts due to my studies and research into natural healing, yet in practice it was not until 10 years ago that I became an advocate for its use personally. For 30 years I held to the Vegan doctrine, except for an occasion when I suffered malaria, and like the Buddha (taking rice-milk), I went into a Samadhi trance due to immediate increase of Ecstatic Regeneration. After many years of misgivings and suffering, lacking health and inspired wisdom in living, near death from pesticide poisoning, finally brought me the illumined insight to disregard the untrue Vegan dogma, allowing the use of clabber (curdled sour milk), and with it likewise came the abandon of depending on Roman Catholic Church dogma for Spiritual inspiration. The result was a remarkable increase in health and Spiritual Insight. Of course, there was another perspective here as to cows, in that unless man provides pasture, cows could not live, and if not used for obtaining milk, they go to the butcher here, besides being raised on pasture that requires no poisonous pesticides in the farming practices. In actual practice, the keeping of a cow and donkey soon becomes an endearing symbiotic relationship, animals becoming "part of one's family".

Dr. John H. Kellogg found that the taint from decaying flesh could be removed by immersion in buttermilk, and no putrefactive bacteria will remain in beefsteak if kept in this sour milk; if it is renewed periodically it will not rot. This basic principle of using the friendly bacteria of cultured milk to eliminate intestinal putrefication, the basic cause of any lack of health, not only was used by Battlecreek Sanitarium, the largest Sanitarium in the world in the 1930s and 1940s, but also the basis of research by Nobel Prize winning Elie Metchnikoff as to health and longevity properties. This use of bacterial plant life for food is what accounts for the fair complexion, hair and blue eyes in European peoples, along with the use of vegetables and fruit as the basic part of their diet, which offsets their use of animal flesh and grains. In turn, one sees such a contrast, for instance, with Andean Indians, Chinese, etc. who since their racial beginnings have not used dairy products and subsisted on grains and legumes almost exclusively, which shows in their short stature, black hair, and characteristic racial color. Yellow color in the skin, if not the whites of one's eyes, shows liver trouble, the yellow or brown bile discoloration and coarseness, if not blemishes coming from toxic waste elimination.

    Dr. Jensen points out the darkening of the Gastro-intestinal area in the iris, the muddy brown or black ring forming close to the pupil, which indicate "great numbers of bacillus coli, a gas producing bacillus, within the intestinal tract and a lack of friendly bacterium, or acidophilus bacillus...Many people eat too much starch and not enough protein, or vice versa, or starches and proteins of low biological value, or do not include enough vegetables and fruits...We have had laboratory fecal tests run by the hundreds, and have found in nearly every case that the bowel harbored about 85% bacillus coli and only 15 per cent acidophilus. These percentages should be reversed...This black coloration (in the iris) represents an acid-alkaline imbalance of the flora, or a toxic condition, and is conducive to the development of worms, bacillus coli, etc. The blood is only as clean as the intestinal tract because it absorbs both nutritive and toxic materials from there." Dr. Jensen visited us at Vilcabamba and studied the centenarians here as he has done in other longevity regions. In regard to this he says, "Large amounts of organic sodium and calcium are essential in the diet to replace that which is burned out by excess acids. When the sodium has been burned out, the calcium is thrown into the joints, as SODIUM, THE YOUTH ELEMENT, KEEPS CALCIUM IN SOLUTION." When the body is "turning to stone", with stiffness, rheumatic conditions, etc. he recommends the use of foods rich in organic sodium, such as celery juice and whey. The whey or liquid which is removed from cottage cheese, butter, and other milk solids contains the health and longevity factor. People who object to dairy products, claiming them mucus forming, would have had little difficulty if they paid more attention to getting the whey liquid, rather than seeking sensual appetite satisfaction in eating only curds. In the case of students who have inadvisably gone on strict fruit diets using market fruits which are usually heavily sprayed or otherwise treated with poisonous chemicals, it has brought on complete loss of health and degeneration of body.

The poisonous chemicals destroy the intestinal flora which digest one's food. In the countries where people live on grains, legumes, and other low quality diets, a great percentage of the children die before adolescence because they cannot develop the needed intestinal flora required to nourish the body and consequently become hosts for disease organisms.

In our present work we are continually referring to dietetic and natural living habits, which make people, tribes and races what they are, because we are sponsoring a New race and New Age of Paradisians free of the factors that caused racial discrimination. It does no good to describe degenerate traits in people, saying it is because of their race which they cannot alter: What can be altered is diet and living habits. Knowing the Truth, we can avoid suffering by not sowing more suffering by living right, purifying our bodies, and teach by being True.

Now, let us study more of the many sources of data that we have as to the Garden of Eden in East and West Turkistan. This interesting clue was presented in "Grace and Race" published by David K. Stacey (Vol. 18, No.4) of New Mexico, U.S.A. "Genesis 4:10: A lake sprang up in Eden to supply the Garden with water, from there it divided and became four rivers...Such a location of 4 rivers starting from one source we find in the PAMIR plateau in Central Asia, between the Tan Shan Mts. on the north, and the Hindu Kush on the south. From the lake of that plateau issue 4 great rivers: the Indus, the Jaxartes, the Oxus and the Tarim. Oxus is called by natives "Dgihun" or Gihon, the central branch of the Indus answers the description of the Pison, the Jaxartes is the original Euphrates and the Tarim going to the East is probably the Hiddekel."

"The Palmirs plateau of today is a different place from what it was 6,000 years ago. At that time the whole of Asia was lower than it is today. At that time a large inland sea covered the steppes of Southern Siberia of which the Caspian and Aral Seas are remnants. Over the frozen steppes of Northern Siberia roamed the mammoth and saber-tooth tiger. Northern Siberia had a semitropical climate and ideal conditions prevailed on the Pamir Plateau. The Alpine lakes watered the Garden of Eden (Pamir) and constituted the headwaters of the 4 rivers."

"Adam in Hebrew means 'to show blood in the face', i.e. to blush, or turn red (Strong Concordance). Thus, Adam and Eve were white and begat white offspring. The Pre-Adamic destruction of cities (becoming without form and void: Jer. 4:23-26) may have included Atlantis and Lemuria. The Adamic white race was and is ordained to bring Light and Order to the Pre-Adamic races (Tree of good and evil in Garden of Eden) but Adam and Eve disobeyed the Divine Command not to mix with other Pre-Adamic races. Eve was seduced by "that old Serpent" called the Devil and Satan (Rev. 12:9). Snakes don't talk so Hebrew "Naw Kash" meant the Enchanter or (black) Magician. Two seeds were planted in Eve, the seed of the "Serpent" and Adam's seed. (And I will put enmity between you and the woman, and between your posterity and her posterity; her posterity shall tread your head under foot, and strike him in the heel, God says to the Serpent in Gen. 3:15). Cain and Abel were womb brothers but of two fathers. Cain was never listed as a son of Adam, Abel was of Adam. Seth was "another seed" of Adam appointed to Eve (Gen. 4:25). Cain killed Abel and the Enmity between the Serpent seed and Adam Israelite seed exists to this day."

Continuing to support their Holy bible, "Grace and Race" seeks to correct things by showing that a black man seduced Eve starting all the trouble in Race-Mixing, without explaining how a black man had become so intimate and acceptable to Eve, or how he had gotten there from either southern India or Africa. I think more reasonable is for them to show that Abel's white race of Aryans were herdsmen, achieving health and longevity by using sour milk products, wild fruits and herbs they encountered in grazing, and that the Chinese were the early inventers of the 5 grains and legumes as Chinese Taoist Scriptures maintain. As the "Secret Doctrine" (Blavatsky) holds, North Asia is as old as the Second (Hyperborean) Race, and "It must be noted that LEMURIA which served as the cradle of the Third Root Race, not only embraced a vast area in the Pacific and Indian Oceans, but extended in the shape of a HORSE-SHOE past Madagascar, round South Africa (then a mere fragment in the process of formation), thru the Atlantic up to Norway." This Secondary Age Northern Lemuria can easily be seen on an under-water Relief Map which shows a submerged elevated continent and ridge or Mt. Range of 9,000 ft. in height within the Basin of the Atlantic Ocean, where it was designated as the continent of ATLANTIS, the Cradle of the Fourth Root Race. "Prof. Berthold Seeman not only accepted the reality of such a mighty continent, but regarded Australia and Europe as formerly portions of one Continent,- thus corroborating the whole 'Horse-shoe' doctrine." Dr. Seeman writing about "Australia and Europe Formerly One Continent" (Popular Science Review, Vol. V, p.18) states, "The singular correspondence of present flora of the Southern U.S. with that of the lignite flora of Europe induces (botanists) to believe that in the Miocene period Europe and America were connected by a land passage, of which Iceland, Madeira, and other Atlantic islands are remnants; that, in fact, the story of an ATLANTIS, which an Egyptian priest told to Solon, is not purely fictitious, but rests on a solid historical basis...Europe of the Eocene Period received the plants which spread over mountains and plains, valleys and river banks (from Asia generally), neither exclusively from the South nor from the East."

But returning to Pamirs, which we were discussing above, if one will follow the 40 degree latitude from Pamir to the Tarim Basin, and then continue on to the Pacific Coast, it also goes thru the Peiping, China, area. Thus, when "Adam was driven Eastward out of Eden (Gen. 3:24) or from Pamir we find the Great Tarim Basin which is one thousand miles wide. "This is the Greatest Sinkhole in the World, altho it is surrounded by the highest mountain peaks in the world, yet its floor lies in many places below the level of the Indian Ocean, indicating that a Great Cataclysm tore the earth here in a Bygone Age. When the Basin floor dropped, the artesian waters (from the surrounding mts.) came forth to FLOOD the land (Gen: 7:10, the Deluge of Noah). The Tarim Basin became a fresh water Sea, and the draining off, created flooding of the Hwang-Ho (Yellow River of China) lasting 3 generations. Thus, in the Chinese "Shu King" we find the Emperor Yaou appealing for one of ability to undertake the controlling of the ever-increasing floods. This plea comes in 2286 B.C., or 58 years after the Deluge of the Bible. The Chinese sacred book of Shu King speaks of Fu-hi, the Chinese Noah. Fu-hi was born of a rainbow and bred and saved seven kinds of animals for sacrifice. The Chinese and Hebraic Bible date correspond exactly. The earth refers to "the land", as used when Atlantean Cain was driven out of the earth, and later the whole "earth" was flooded means that the land of the Tarim Basin was flooded so high mountains (meaning high hills only) were covered 15 cubits. (25 to 31 ft.) in the Biblic Flood. Arat means "tops of hills", the location in Armenia being unknown till recently, formerly being called Mt. Massis, this being where Noah moved in 2344 B.C. In Ur of Chaldeans the high hills of Tarim Basin were converted into Arhat of Armenia in the retold legends of the Hebrews. These facts of the "Grace and Race" treatise, along with those about Moses obtaining Ethiopian-African wives for himself and his people, would better indicate that Cain's wife was of Chinese or Mongoloid characteristics, rather than black. As the Bible tells it, man first was cast out of warm Paradisian Hyperborea so he had to wear animal skins to keep warm when the earth's axis changed the climate, and thus to survive he had to nourish from the milk of animals that were protected from freezing and provided by chance with vegetation humans found palatable. The breeding of the 5 grains and legume foods, from small grass seeds occurred about 10 to 15 thousand years ago, so that the last remnants of Atlantis of 11,000 years ago had become a Mongoloid race from liver discoloration of skin, which correspond to the legend of Cain, or Dragon-Child of Chinese Legend and Serpent-Seed of the Bible.

In the first 4 chapters of the Holy Bible, the Divinity is called Elohim, but from the 4th Chapter the Hebraic origins begin to call God Jehova (or Yahveh). Verse 1 in the original Hebrew, as well as English versions, reads, "Adam knew Eve his wife and she conceived, and bore Cain, and said, I have gotten a man for the Lord. The significance is that Jehova and Cain are identical, confirmed also by the Rabbins teaching that "Kin (Cain), the Evil, was the Son of Eve by Samael, the devil who took Adam's place"; and the Talmud adds that "the evil Spirit, Satan, and Samael, the Angel of Death, are the same". Correctly translated the last verse (of Chapt. 4) reads, "then men began to call themselves Ja-Hovah", Jah meaning male, and Hovah (Heva or Eve) female. Before this the race was sexless, both male and female, as described of the Heavenly man, Adam, and when Eve was made of a rib, bone of the bones, it means the previous races were boneless, without need of a skeleton. Satan represents the REVERSE OR POLAR OPPOSITE OF EVERYTHING IN NATURE, or the Adversary, and as the Gnostics held, Jehova was the Creator of the Serpent, Satan or Evil. They called this god of Moses, Ilda-Baoth, or Demiurge. (see Codex Nazaraeus, on Fallen Angels) As it says in Psalms 82:6, "I have said, You are gods; all of you are children of the most High." "I am that I am" is what Jehova says his name means, and thus is the ever living male-and-female principle. (Ex. 3:14)

As the SECRET DOCTRINE Vol. II reveals: "If, then the teaching is understood correctly, the first continent which came into existence capped over the whole North Pole like one unbroken crust, and remains so to this day, beyond that inland sea which seemed like an unreachable mirage to the few arctic travelers who perceived. (Tales of the Hollow Earth, astral visions of past cultures). During the Second Race more land emerged from under the waters as a continuation of the "head" from the neck. This shows that Northern Asia is as old as the Second Race... Asia is contemporary with man... Between the first and second races the eternal central land was divided with the water of life. It flows around and animates mother earth's body. It's one and issues from her head (cap of Hyperborea); it becomes foul at her feet (the South Pole). It gets purified on return to her heart, which beats under foot of the sacred SHAMBHALA which then was not yet born. For it is the belt of man's dwelling on earth that lies concealed the health and life of all that lives and breathes. During the first and second races the belt was covered with the great waters. The great mother travailed under the waves and new land was joined to the first one which our wise men call the head gear (cap). She travailed harder for the third race, and her waist and navel appeared above the water. It was the belt, the sacred Himavat (mt.) that stretches around the world. She broke toward the setting sun from her neck downward (to the S.W.) into many lands and islands, but the eternal land (the cap) broke not asunder. Dry lands covered the face of the silent waters to the 4 sides of the world. All perished in turn. Then appeared the abode of the wicked (the Atlantis). The eternal land was now hid, for the waters became solid (frozen) under the breath of the nostrils and the evil winds of the Dragon's mouth/." From this quoted material, H.P. Blavatsky continues explaining from various ancient and modern texts in geologic and other scientific findings, how the horse-shoe continent circled the earth as described earlier.

To understand the Gnostic Teaching of John in his Gospel and his Apocalypse, we must remember that the Gnostics abandoned the Pharisee and other popular Jewish doctrines, reforming even the Essene tenets, for a new Gospel. "To him that overcometh I shall give to eat of the tree of Life...To him that overcometh shall be clothed in white...Such as I live, I rebuke and chastise. Be zealous therefore and do penance... These are they who are come out of great tribulation, and have washed their robes (raiment or body), and have made them WHITE IN THE BLOOD OF THE LAMB." As we said earlier the ethnic or racial color of people is determined by how purified their racial and personal diet is. In the last supper scene of popular N.T. translations, Jesus says, "I shall not partake of the fruit of the vine, until that day when I shall partake of it fresh (new) in the Kingdom of God" (Paradise Regained), after he states the Holy Grail containing the fruit of the vine, grapes, is his blood, and clenching the fruit he sheds his blood (grape-juice) sharing with the Elect of his body and blood thru which we become white in raiment. Thus, John writes, "Blessed are they that wash their robes in the blood of the Lamb (the Savior in paradise and not Jewish sacrificial lamb), that they may have a right to the Tree of Life (Everlasting) and may enter the City (of God)." That this is directed against "them that say they are Jews (Hebrew-Ebrons in Aramaic) and are not, but are the synagogue of Satan," when the allegorical truth is translated. Ebrons comes from Heber meaning To Cross Over, or Overcome the purely material and pass over to higher God Consciousness. Rome was in conflict with the Jews so the Roman Church refused to translate the full Gnostic or Esoteric meaning of the Gospel allegories. Worse, Rome, like the Jews, the old Babylonians and Atlantians all had become corrupt world conquerors, fed on sacrificed beasts besides engaging in racial adultery. In blessing the tribe of Judah, or the Jews, Jacob gave a key to the Apocalyptic allegories: "he hath purified his raiment in wine (meaning living water of Paradisian fruits), And his robe in the blood of grapes, His eyes shall be red (reflect pure Spirit, in original Hebraic symbol), And his teeth white with milk." The 12 tribes were to gather together in Hebron meaning Promised Land, which Gnostics interpreted as God's Paradisian Kingdom. However, instead the Jews, "He that shall lead into captivity, shall go into captivity, and he that shall kill by the sword, shall die by the sword," as John said, showing their Egyptian and Babylonian bondage, and violence, beside corruption, made them a synagogue of Satan.

By comparing the Mystery of Babylon the Great, with Eastern esoteric teachings, and the teaching of the Great White Brotherhood of the 144,000 Elect that Apocalypse continually refers to as being the purified white blood or race, we find the same Seven Root Races, and Ages or Cultural Origins. "I tell thee the Mystery of the woman and the beast that carrieth her, which hath 7 heads and 10 horns... The 7 heads are 7 mountains (lands, continents and races thereof), upon which the woman sitteth, and they are 7 kings: Five are fallen, one is and the other is yet to come..." This includes (1) the Ethereal Race of the Immortals, (2) the Hyperboreans, (3) the Lemurians, (4) the Atlantians, (5) the Aryans (which began with Krishna), (6) the Era of Christ (from Roman Church Empire to Merchant's (Capitalist) Empire, and (7) the New Race of Paradisians, God's Kingdom, of the prophesied Maitreya in South America. "Because all the nations have drunk of her wine of wrath of her FORNICATION (racial defilement, eating forbidden flesh, seeds, morals) and the kings of the earth have committed fornication with his Harlot, and merchants of the earth have been made rich by the power of her delicacies." The Book of Life of this earth is sealed with 7 seals, so by opening the first seal, the Spiritual or White Race is revealed, the 2nd Seal reveals the Red Race, the 3rd Seal reveals the Black Race, the 4th Seal reveals a Pale Race, the 5th Seal reveals those of White Robes who died for the sake of the Word of God, Camp of Saints, and the 6th Seal shall reveal those freed from bondsmen and merchants who shall make the earth green again, while the 7th Seal opens up the new Golden Age and Paradisian Race. But why was Babylon the MOTHER OF HARLOTS AND ABOMINATIONS? Nimrod, the mighty warrior before the Lord was an ATLANTIAN.

An Arabian Historian, Masoudi, tells of the Nabatheans, believed to be Sabean star-worshippers of Chaldea, who told of their origin thus: "After the Deluge, (the sinking of Atlantis?) the nations established themselves in various countries. Among these were the Nabatheans, who founded the city of Babylon, and were the descendants of Ham, who settled in the same province under the leadership of Nimrod, the son of Cush, who was the son of Ham, great-grandson of Noah... This took place at the time when Nimrod received the governorship of Babylonia as the delegate of Dzahhak named Biurasp." The translator, Chwolsohn, under the non-conspicuous title, "Nabathean Agriculture" was thus able to present a "complete initiation into the mysteries of Pre-Adamite nations, on the authority of undeniable authentic documents". The work is a re-translation from Arabic, into which it had been translated from Chaldean, brought to European publication in 1860 by Chwolsohn, who placed the Arabic version as having been made in the thirteenth century B.C. In the first page, it states, "the doctrines propounded therein, were originally told by Saturn to the Moon, who communicated them to her idol, which idol revealed them to her devotee, the author, Qu-tamy" who was a wealthy Babylonian landowner. Of course, various writers have denied its antiquity, since to speak of Noah and Nimrod, it places this with Moses writings which were first published only in the 3rd Century B.C. by Esdras. Nebo, or Nabo in Greek, of the Nabatheans was the deity of the planet Mercury, the god of Esoteric Wisdom, or the Greek Hermes, and Thoth of Egypt. This means the Nabathean were an Esoteric Brotherhood of the Sabean Astrologists, which (along with Chaldeans and Egyptians) the laws of Moses were most severely enacted. Furthermore, various writers hold that Noah was an Atlantian, a Titan, meaning a giant, which was overlooked purposely in the Hebraic Genesis. M. Bailly is among those who depreciate Biblic chronology, seeking to prove that the very ancient northern nation of Atlantians existed long prior to the Hindus, the Phoenicians and the Egyptians. In turn, "Atlantis, the Antediluvian World", by Donnelly claims that the roots of today's institutions reach back to the Miocene Age, and from the Miocene Atlantians, the Aryan colonies received their arts and sciences. (Miocene= 28 to 16 million years ago in recent estimates, altho these estimates were of 7 million.)

In summary, from studies made by Bunsen and other

researchers, P. B. Randolph in "Pre-Adamite Man" dated Nimrod as having existed 16,000 B.C., the Flood in Asia and part of Europe about 21,000 B.C., Founding of Thinite 1st and 2nd Dynasties in Egypt 28,000 B.C. Randolph himself in turn held that not less than 42,000, nor more than 58,600 years ago that the most tremendous event this earth has ever witnessed happened when the earth changed its axis, the sun melted the snows at former poles of the earth, and the tropical plants and animals now found in the Arctic regions became frozen solid.

In the liberal "Historians History of the World" (H. Smith Williams tells us: "We learn from evidence graven in stone and baked indelibly in bricks that in the year 4004 B.C. which our Bible margins point out as the year of Creation, vast communities of people...had attained a high degree of civilization...And from that day to this stretches the thread of civilization, UNBROKEN BY ANY UNIVERSAL FLOOD OR CATACLYSM." Herodotus had colored history with what he saw in the temples of Babylon where he alleges, all women of the city, of whatever class or rank were obliged to prostitute themselves for hire. However, as Prof. H. G. Wunderlich wrote in "Secret of Krete", "The children of Gods were not the offspring of regular marriage, but were born in a temple of a God (frequently Zeus or Apollo) because the mother did not have a house of Grecian Gods, but already in Babylon girls received pre-marital training, having a child with the God of the Temple, considered a religious duty which also brought in Temple revenue needed for upkeep from young men seeking sexual experience. Some women never found husbands but served the temple. Babylonian history is filled with wars, conquest, gory cruelty, sensual lust and luxury of their kings, and reflects the fact that already the Sumerians enjoyed pork and pork-fat, beef and veal, raised grains and legumes, beside fruits and vegetables, cheese, etc.

Cicero was skeptical about Babylonian monuments preserving astronomical observations dating over a period of 270,000 years. The Sumerian King List describes eight reigns that span a quarter million years. Dodorus tells us Assyrian history was already accepted for 1800 yrs.

Ninus is the First Native King of Assyria recorded in history for his "many great and noble deeds". With the Prince of Arabia, with their great armies they invaded Babylonia destroying their monarchy to impose a yearly tribute upon its inhabitants. Then he invaded Armenia, where King Barzanus met him with presents, but was forced to give him the kingdom of Armenia, being unable to deal with Ninus. Being strengthened, he invaded Media, who King Pharnus coming out with a mighty army was utterly routed, and taken prisoner with his wife and 7 children and afterwards CRUCIFIED. He then marched against all the other provinces of Asia and subdued them, all except the Indians and the Bactrians. In the end he subdued the Bactrians also after a long and tedious siege.

Then King Ninus married Semiramis. In Ascalon, Syria, is the temple dedicated to Derceto (Dagon) who was a woman in face and fish in other parts of her body. Venus caused her to fall in love with a handsome young man, and ashamed of a child that resulted, she killed the man and exposed the girl-child on the bank of a deep Lake, and cast herself in the lake and was transformed into a fish. To this day Syrians eat no fish, adoring them as gods because of this, says Diodorus. The infant was fed by pigeons who took milk obtained from shepherds to nurse her, and as she grew older brought her pieces of cheese. Then Shepherds discovered where their cheese went, and adopted the child, naming her for the pigeons, Semiramis. A Syrian nobleman carried her to Nineve, a city just founded by Ninus, and after two children with the nobleman, she was taken by Ninus due to her beauty and valor when she helped him invade a Bactrian city by scaling a wall under cover of night.

Semiramis became Queen Regent of the Empire after Ninus died. She then built the great City of Babylon, with the help of 2 million workmen, with the Euphrates River running thru the middle of it. With two Palaces, one on each side of the River, a bridge connected them. There were all kinds of statutes of Ninus and Semiramis in hunting scenes with all sorts of wild beasts, which makes one wonder if the author of the Nimrod, mighty hunter before the Lord, Bible story had not been influenced by this visual record of history, and Ninus was the "founder of Nineve and Babylon" (thru his wife.) She built temples in other parts of the Empire, also invaded Egypt and Ethiopia, and finally sought to conquer, India, seeking to scare the elephant mounted Hindus with a gigantic Mock Elephant, but failed. Perhaps here we have the prototype of the Great Harlot drunk with blood of Saints, fornication and pleasures.

CHAPTER V: THE EGYPTIAN INITIATION INTO THE ATLANTEAN MYSTERIES, THEIR SACRED HIERATIC LANGUAGE, AND TRAITS.

"You do not know which was the best and most handsome generation of men which has ever lived on this earth. Only a weak seed of it, of which you (Greeks) are the descendants, is all that remains. Their books preserved the records of a great nation, which emerging from the Atlantic sea had invaded Europe and Asia. The Greeks were but the dwarfed and weak remnant of that once glorious nation." Thus wrote Plato in his "Timaeus and Critias", Critias being Plato's 90 year old grandfather who as a child learned it from one of the Grecian Seven Sages, Solon, who in turn was taught by the Sages of Egypt, the priests of Sais.

"First of all, one must remember that 9,000 years have elapsed since the WAR OF THE NATIONS, which lived above and outside the Pillars of Hercules, and those which peopled the lands on this side," Plato adds. Theopompus describes them in "Meropis" as consisting of two races, a fighting, warrior race of the belligerent, and a pious, meditative race continually visited by the gods. H. P. B. suggests that this former pious, meditative race were the last Atlantian giants, or what the Bible calls "the primitive and mighty ones, prototypes of Nimrod." That the standing army of Atlantis had over a million men, its navy 1,200 ships and 240,000 men, shows that Plato was not describing a small island, which may have sank only 11,000 years ago, but this was a part of a vast continent. In fact, Atlantis was part of a horseshoe continent that we described earlier as having extended south thru Africa to the East Indies to include Lemuria before that part sank ages ago. Back in most ancient time all the Sahara desert was a sea, from which emerged a continent that became as fertile as the famed Nile Delta, which again was submerged temporarily, before it has become the wild desert of today.

The Atlantes, a people of Western Africa which gave its name continent, or an island only 350 by 200 miles similar to the size of Mt. Atlas, were vegetarians, and "whose sleep was never disturbed by dreams and who daily cursed the sun at its rising and his setting because his excessive heat scorched and tormented them", Herodotus, the first historian wrote. Far from being a small Ireland, which so often is mistaken for its full size, Plato's account describes the whole Atlantian continent as being as vast as "ALL ASIA AND LYBIA" put together. If one will take the "ATLAS" (name also derived from Atlantians) or a map of northwest Africa, one can readily locate the Atlas Mts. in Morocco and Algeria, as well as the Canary and Madeira islands. The ancient continent had more mountains than valleys, and the Mountain of the gods, Atlas, was 3 times as high at that time, so that Libyans called it the Pillar of Heaven. The Ancient Poets tried to immortalize the history and legends of the Atlantians by developing allegorical tales of gods. So Atlas became a son of an ocean nymph and his daughter Calypso, "the watery deep under which the progeny sleep. Atlas thus becomes a colossal giant upon which contemporary continents are supported, altho the ancients claimed the feet of the giant tread the earth, while his shoulders support the celestial vault. The shoulders of Atlas, like the Pillar of Heaven, refer really to Mt. Atlas, Tenerife Peak and other huge mountains of Atlantean and Lemurian times. The Atlantes claimed Uranos for their first king, and Plato starts his story of Atlantis by the division of the great continent by Neptune, the grandson of Uranos. This can be accounted for in that Uranus ruled the Second Race and their Continent, while Kronos or Saturn governed the third race or Lemurians, while Jupiter, Neptune and others fought in the allegory for Atlantis, according to Mdme. Blavatsky. Thus, Heavenly Rulers in spiritual interpretation were the Dhyana Chohans, or the seven great Leaders of seven great ages, races and continents. The confusion as to the existence of Atlantis comes about because each language, and author of old translated the original Atlantian names into meaningful names of their own language, or left them in obscure forms of translation, not readily recognizable.

Referring to the last small island of the Great Atlantean Continent or "Poseidonis", also of Greek legends as name for Neptune, Plato wrote that the plain surrounding the city was itself surrounded by mountain chains. The plain was smooth and level, and of an oblong shape, lying north and south, 3000 stadia long and 2000 wide. They surrounded the plain by an enormous canal or dike, 101 feet deep and 606 feet broad, and 1,250 miles long.

We have spoken about the invention of the 5 grains which brought about racial type degeneration, suggesting the Tarim Basin and East Asia as the origin. Chuang Tzu claimed that the Spiritual Man whose skin was white as snow, did not eat the 5 grains, which have been listed by some as rice, millet, barley, beans and wheat. The remnants of the Atlantians were yellow, red, brown and black, and were dwarfed in size. The reduction of racial stature is pronounced here in the high Andes where the Quichua Indians live on corn and barley almost exclusively, using almost no green leafy vegetables and milk products. Their racial features in color, the architecture of American Pyramids, and so many traits coincide with early Egyptian culture, so as to seem to indicate similar continental origins as once being part of the great Atlantian continent, and more certainly their people.

In the Book of the Dead, the Egyptian Isis states, "I am Queen of these regions: I was the first to reveal to mortals the mystery of wheat and corn. Rejoice, O Egypt, thou who wert my nurse." Wheat was so sacred that it was buried with the mummies of Egyptian priests. It was when mankind abandoned the guidance of Gods, the collective Logos, being left to their own resources and industry, that inventors appeared who discovered the use of fire, wheat and wine, who in turn were made gods in gratitude from their people. Herodotus gave an interesting report: "Wheat and barley are common articles of food in other lands, but in Egypt they are thought mean and disgraceful. The diet here consists principally of spelt, a kind of corn which some call zea. Male children are left in other nations as nature found them, but in Egypt they are circumcised...They have two sorts of letters, one which is appropriated to sacred subjects, the other used on common occasions (the hieroglyphic and hieratic characters). They are so regardful of neatness, that they wear only linen and that always newly washed. The priesthood wear one garment of linen and their shoes are made of byblus (papyrus), they wash themselves in cold water twice a day and as often at night, and every third day shave every part of their bodies, to prevent vermin or any species of impurity from adhering to those who are in service of the gods...Beans are sown in no part of Egypt, neither will the inhabitants eat them either boiled or raw, the priests will not even look at this pulse esteeming it exceedingly unclean. The Egyptians venerate cows beyond all other types of cattle. The Egyptians regard the hog as an unclean animal, and if they casually touch one they immediately plunge themselves, clothes and all, into water. Marriage with swineherds is studiously avoided, so they are reduced to the necessity of intermarrying among their own profession." We have purposefully avoided mention that the Egyptians did eat meat without blemish, sacred bulls, etc. However, my point is to bring out that the Essenes got many of their practices from a similar source, Pythagoras, who studied in Egypt, also held beans to be an unclean food, and that the Hebrew language, customs against use of hogs for food and special clean sacrificial animals, beside circumcision were of Egyptian origin. As described earlier, the Sumerians, Babylonians, etc. did not have this disdain for pork, etc.

Thru many years of dietetic research, we have found such findings most interesting. It is evident that the early Egyptian Hierophants were the product of migration from Atlantis of the Spiritual Aryans who were versed in the Higher Sacred Sciences, as taught by Thoth the Atlantian (called Hermes by Greeks), while the swine-eating Sumerians and Babylonians were the Mighty Warrior race descendants characterized by Semiramis. However, just as the Atlantians had degenerated into racial types by dietetic impurity, likewise the pure and noble Egyptian culture reached a point of decay, first eating "meat without blemish" and gradually drawn to a more sensuous diet and racial mixing, the negroid or Lemurian element became integrated in the race. About 688 B.C. we learn of Teharka, son of a Nubian woman, became the first mulatto to ascend to the throne of Egypt, as prophesied by Isaiah, and "So shall the King of Assyria lead away Egyptian prisoners, and the Ethiopian captives, young and old, naked and barefoot, with their buttocks uncovered, to the shame of Egypt," (Isa. 20:4), just as Herodotus noted in the 5th century B.C. that Egyptians ate beef, and even pork in celebration of Bacchus, or made a pig-shaped figure for the sacrifice from baked bread if they were poor. The Atlantian vegetarian noblemen who had once held fish, flesh and fowl, barley, wheat and beans as impure foods, thus became infiltrated with impure gods, foods and traits.

In my study of the fruit diet, I have had a very difficult time determining if avocados were not the first tempters of the cunning old serpent of Eden. They certainly are a fruit, contain a seed yielded for their reproduction, are comparatively low in protein (1.7%) and have a 75% water content, but alas, they also have a 26.4% FAT CONTENT that is EQUAL TO THE FAT CONTENT OF PORK CHOPS and ham! Already, while still living in Florida, in 1939 and on a strict fruit diet without vegetables I had found that an exclusive avocado diet for one day brought copious involuntary seminal losses at night during sleep, so evidently it was not prescribed by God in the Garden of Eden to be eaten freely of, like other juicy fruits. The following year, the first place I lived at in Valencia of Quevedo, Ecuador, was a cacao (cocoa-bean) grove, which had avocado trees planted alternately for fattening hogs for pork and their service of rooting out weed growths around the cacao. However, contrary to gaining any features of the fat hogs, due to an amoebic dysentery, my 6 ft. 4 inch frame went down to 97 lbs., worse than even on my experiments trying to live without eating later for 6 month spells, altho avocados were ideal for hog-raising. This association, along with resemblance to pork-fat that most tropical avocados have, influenced me to shun caring for avocado trees, altho I have enjoyed the eating of all kinds of other fruits that I had planted in various Paradises I started, forever thinking that when I have papayas and other fruits producing, avocados would play an evil part in Paradise. Yet, in giving up cooked foods and salad oil as a health measure, only clabber and avocados were left as essential ingredients to dress the greens or vegetables in salads. It was very hard to get used to the idea of using soured milk transformed into acidophilus bacterial plant culture, after 30 years abstinence due to a Vegan philosophical ideal I learned of in a magazine. But near death from pesticide poisoning of the intestinal flora, evacuating blood,- beside later healing of a tumor, is how I found the very thing I had condemned, nourished my body back to health. Of course, unless the milk used to make clabber comes from cows that are healthy, raised on wild grass and leaves free of poisonous farm chemicals, and not treated with antibiotics or pesticides, as was my case, this would not be valid. Moreover, in U.S.A. and most affluent countries, herbivorous cows are overfed on grains, legumes and other concentrated seed meals to greatly increase milk production, so milk is no longer the product of the green grass of the wilderness, but instead such milk is the product of a mature seed, or a reproductive substance. Cotton seed, corn or other seed milk is not milk made of grass. This is why U.S.A. school children suffer mucous conditions, tooth decay, and all kinds of childhood ailments altho they may have plenty of milk products that came from cows fed on grains beside grass, as well as eating grains and other reproductive substances that require disease microbes as scavengers. Grains are likewise used to stimulate an increase in egg production, but no one needs more reproductive substance, since giving birth subtracts health and needed elements from the mother.

The long-lived grazing people that migrated thru the wilds from East Asia to West Europe, nourished basically from milk products, which nature designed for the maintenance of health in pre-puberty status of the young (without reproductive needs). However, it was under all these precautions in not violating the health, purity and integrity of cows that nourished exclusively on grass and leaves, free of modern toxic chemical contamination, that the Aryans, or Noblemen of Egypt, India and Nordic nations considered cows and their milk sacred or a gift of the Gods, and Paradise is described as a Land flowing with milk and honey. Likewise in our mission of Paradise Building, we find it unhealthy to locate near farms which have an avocado, milk or fruit supply, especially indicative of toxic chemical contamination, and since the wilds provide little but wild greens, the cow and donkey thus become necessary companions for man's survival until one's garden and orchard produce. Flora is needed to digest grass or greens.

Not only does the excess fat content of avocados, similar to pork, cause reproductive losses when used as a diet, but our research in recent years has shown that excessive use in proportion to greens that compose a salad, will produce smegma elimination in males and ovulation in females, beside liver hardship. Thus, it is similar to the Egyptian and Essene observations recorded in the Bible in which the eating of pork was considered unclean, or produced Smegma and inflammation of the glans, just as the uncircumcised were viewed as being "unclean". "Between the propuce and the glans of the uncircumcised boy, or man, is secreted a tallow like substance called Smegma. Left uncleansed this may cause inflammation and trouble. It is very likely to stimulate sexual arousal and lead to masturbation. This is the chief cause of serious sex problems in countless young men before marriage," observes Herbert Armstrong. Circumcision was thus given as Gods commandment to the Jews and the Egyptians, while even now it is medically recommended to avoid trouble, just like removing tonsils, adenoids, appendix, hair, etc. enables people to fit into degenerate life styles, rather than regenerate man with healthy lifestyles.

In the September 1978 DR. SHELTON'S HYGIENIC REVIEW, Dr. V. V. Vetrano scientifically illustrated that menstruation is a pathological condition. "A woman who menstruates is not necessarily a woman who can bear children...Ovulation and menstruation are separate and distinct phenomena...Hygienically speaking it is a sign of health if the woman menstruates very little or does not menstruate at all if her general health is excellent...leukocytes in tremendous quantity are released during menstruation just as they are in pathological hemorrhage." In this connection, I might add that the consumption of any cooked food also increases leukocytes, which is called "the phagocytic index", and this includes cooked, distilled water, causing both enervation and expenditure of vital substances needed to produce the excessive blood cell needs. This augmentation of the phagocytic index two or three times daily with the use of cooked food seems directly related to a climatic need of discharging excessive reserves of blood cell waste, accounting for menstruation, ovulation, seminal losses, etc. in a monthly rhythm. Alexis Carrel and other eminent scientists have found no reason why humans should die if conditions of hygiene and cell nourishment are met. If a very small percentage of humans ceased to die, for lack of earth resources (land, water, etc.) this same law would require that the immortals would not reproduce, since replacements would destroy the integrity in demographic excesses. The only reason man harbors a reproductive instinct, and the cells of the body prepare for death, is because the reproductive enzymes taken in food and pathological toxins ingested, program the body to age and die.

Dr. Shelton added these observations of interest to our study of the Vitalogical Sciences,- "It is a fact that some women in civilized life never menstruate, and yet they bear healthy, well developed children...We know that girls may and do conceive before they ever menstruate and women have conceived after they have ceased to menstruate. The process of menstruation is not a physiological one, being the result of an interference with nature, of a thwarting of her laws. Menstruation is a hemorrhage, being attended with a rupture, and no hemorrhage is natural. Hindu women as a rule do not menstruate. With them menstruation is considered a crime. History does not furnish unequivocal evidence that menstruation was common in ancient time." (The last 3 sentences he quotes Dr. King.)

Dr. Shelton's requirements for a fluxless ovulation (without menses) include physical and special exercise, sunshine, rest and "A healthful diet of fruits and vegetables, largely if not wholly raw". Dr. Vetrano adds to his formula avoidance of all animal products and chemicals of all kinds, including insecticides, etc. Now, the question remains, why is it that Hindu women as a rule do not menstruate, as both of them quote as examples, and yet they are known to hold the cow sacred, freely use curds, milk, polished rice (without germ) etc. Our findings show no reason to hold such a fetish in avoiding acidophilus bacterial plants. So, in turn, we advocate avoiding seeds, nuts,- which they use freely,- because such reproductive foods nourish the greatest survival problem in creating excessive population here and now. Child-bearing acknowledges death, and if our fight for survival is hopeless, world conditions for children will be even worse evils. Only when there has been great abstinence from giving birth by the cultural leaders or regenerate, has caution become general among all.

Now, it has been objected that the human body cannot assimilate colorless plant life, that is lacking chlorophyll, such as mushrooms, yeast, and thus would include bean and other seed sprouts, bleached celery, lettuce, cabbage, etc. beside acidophilus Bacillus. We are in accord, especially as to yeast, mushrooms, sprouts, etc., and see how they can cause mineral imbalance as claimed, rather than provide minerals, altho this would not be the case if the diet was bad and worse before in comparison.

However, we are not desirous or want to assimilate or metabolize the acidophilus plants in clabber, but instead the object is to restore them to the intestinal tract. As Elie Metchnikoff has shown, man comes into the world with his intestinal contents free from any form of germ life, but in a few hours microbes penetrate, and henceforth the intestinal microbes are determined by the food we eat. When mother's milk, or other milk of herbivorous mammals is introduced in the intestines, harmful Bacillus coli are replaced by useful Bacillus acidophilus. This friendly flora makes possible the digestion of herbivorous foods, greens, salads, etc. in humans, just as calves would die of starvation if they never got the milk required to start their Lactobacillus flora. Most adults lose 85% of lactobacillus, and of late most strict fruitarians and health conscious destroy it by use of poisonous market fruits and vegetables unaware of the danger, while still more destroy the friendly flora by the use of drugs. Only with a friendly flora do we have the digestive enzymes needed for fruit and salad assimilation and health.

Still another objection comes from those who have found traces of Bacillus acidophilus in soybean and other seed cheese, and other products that do not come from animals like cow's milk. Yet seed-milk was not introduced by those who want to eliminate menstruation, ovulation, seminal losses and smegma by abstaining from reproductive substance, but it is the food of the Chinese and other non-milk using races who rely on soybeans, rice, and other seed foods. In turn the European and other milk-using people lack the intestinal flora to assimilate seaweed, soybeans, etc. provided first in maternal milk.

Those with Vitalogical objectives are not interested in greater seed or reproductive attributes, but instead seek an intestinal flora that will enable one to rely exclusively on fruits and vegetables, which requires the herbivorous type of Lactobacillus. Thus, one will note the obnoxious smell of grain-eating chicken leavings, compared to the herbivorous animal dung. The same observation becomes even more contrasting concerning the menstruation, ovulation, seminal losses and smegma, beside perspiration and body odor (or aura) of those observing the strictest Vitalogical Discipline by absence, compared with people who use seeds, or cooked foods, or flesh and combinations of such foods, and worst of all in alcoholics, drug-users, etc., where the malodorous fetor, ammonia and bacillus coli toxins predominate. The herbivorous and frugivorous acidophilus bacillus require a low-protein, low-fat, 80% water medium to live in, but the seed-decomposing acidophilus require the concentrated seed-protein medium, and with cooked seed and flesh foods, the bacillus coli make it their home.

Another interesting aspect discussed with Iridology By Dr. B. Jensen has to do with Polarity Relations to the Body. Yogic Breathing thru the right nostril, as well as high altitude, make the body more electro-positive, and the opposite make the body more negative. Asthma, toxic liver, nerve damage by sprays, etc. show that an electro-positive diet is needed to charge the right side of the body, beside high altitude and right nostril breathing. Acidophilus milk is recommended as an electro-positive food, beside electro-neutral foods which include all acid fruits and leafy, non-starchy vegetables. Above the ground vegetables are more positive and below the earth surface or root vegetables are non-solar or electro-negative. Correspondingly in Yoga, milk and fruits are recommended for positive spiritual endeavor. To be avoided for an electro-positive diet are avocados, bananas, all fats or oils, sweet fruits, starchy vegetables, grains, legumes, etc. Thus, body building enzyme-proteins concentrated above the ground are electro-positive, while the high calorie energy foods are negative.

This research has been, and may personally be, confirmed by checking observations of the iris, while purifying and healing the body. However, in my own case I found that the strength of my arms increased until I could lift my own body weight 30 times in morning chinning exercise, the ability of which was lost with a few months season of abundant avocados to replace clabber in salads, showing also the replacing of muscle with fat in the body. The protein in tropical avocados was incomplete in muscle building requirements, and like the olive oil I used in earlier Vitarian experiments, the avocado fat put a strain on the liver which was overloaded by toxic waste elimination. In California and Florida, in spite of health ideals and morals against slaughter, like vegetarians unwittingly do today still, we were using avocados nourished on blood meal from slaughterhouses which is dried for fertilizer used by most all avocado-growers. We gave up flesh and food produced by animals, yet the patronizing of slaughterhouse nourishment was supported, as well as the toxic chemical industry that destroys human life, when others grow one's food. As fast as plant pesticides have begun to prevail even in the most isolated regions of the earth, in this mass genocide of World War III in civil conflict of man against man, drugging and poisoning has become the common legalized weapon. Veterinary Medicine also has made vaccination, containing antibiotics beside, obligatory in commercial dairies, and even a small farmer uses pesticides to eliminate ticks, etc. on cows, so pure milk becomes a problem, altho in the case of goat milk such practices are not in use as yet.

In spite of our sensual taste preference of avocados over the use of clabber in salads, making our research biast, yet so far we have not found any variety of fat-less avocado, or other fruit or vegetable to dress raw salads that gave the healing and strength that compares with acidophilus bacterial plants. We know many of our readers object to using animals to produce food,- even if cows or goats thus can live longer lives rather than be slaughtered early in life,- and thus it damages the popularity of our teaching, since people are more receptive to false hopes and unproven impractical idealism, often at the cost of health and life, rather than remain steadfast to the proven living Truth. Nor are our findings now fully in practice in spite of what we observe, since bananas, avocados and other electro-negative foods predominate here in season or all year in our tropical highlands contrary to the regions of most of our readers; and the electro-neutral and electro-positive foods are more difficult to grow or obtain here, in contrast to colder temperate climates.

Now, let us study more on the "EGYPTIAN INITIATION" by Iamblichus or the Teaching of Thoth. "Iamblichus, founder of the Syrian school of Neo-Platonism, was born A.D. 250 at Chalcis in Coele, Syria and died about 330, during the reign of Constantine the Great, emperor of Rome... The Egyptian Initiation is a dramatization of human life as experienced by each person as they journey along the path to the eternal hereafter. Each person enters this life entirely ignorant of what lies ahead but having implicit Faith that they are welcome, so they accept the decisions of their parents as divine law. After childhood when we start life for ourselves, each one is required to continually decide what path to follow to obtain the desires he is striving to accomplish. In this position we are ignorant of the future, never-the-less each decision we make determines the lessons and experiences we will encounter in life as we grow older" (introduction by H.O. Wagner) "We are obliged to learn that Divine Law is inflexible, and having once decided on a plan of action, we must accept what comes. Implicit Faith in a better future urges us onward to complete our destiny, as F-A-I-T-H is the true compass of the soul. The Egyptian Initiation was a gradual training of the whole human being to the lofty heights of the Spirit." Thus, Mr. Wagner gives the Eternal Rule: "ONLY HE WHO RULES HIMSELF IS ENTITLED TO RULE OTHERS". Then, he adds history:

"Many ages before Abraham and his herdsmen wandered over the desert of Arabia, mystical learning from the magical schools of the lost ATLANTIS was carried by tidal migrations eastward toward the rising sun and transplanted in the valley of the Nile. And there it flourished for ages under the fostering care of a mighty priesthood and a colossal sacerdotalism. It was there that HERMES, the first initiator of Egypt, held sway as the Mightiest Hierophant the world has ever known. The Greek disciples called him Hermes Trismegistus, or Thrice-Greatest, because he was looked upon as King, Lawgiver and Priest. About 2,200 B.C. Egypt underwent the formidable crisis of foreign invasion and semi-conquest caused by the great religious schism in Asia which stirred up the masses by sowing dissention in the temples. Led by king-shepherds, called Hyksos, the deluge rolled over the delta and central Egypt followed by a corrupt civilization, Ionic effeminacy, Asiatic luxury, harems and gross idolatry."

The priesthood outwardly bowed before the invasion, seeming to accept the worship of Apis or law of the bull-god. Within their holy temples they concealed the Sacred Sciences. The Mysteries were withheld from all those who were not able to prove their right to receive. "The Lips are closed except to the ears of understanding," was the wise admonition of Master Hermes. The Initiation of Iamblichus is believed to be in use in the time of Ramses, about the time when Moses and Orpheus were living,- the year 1500 B.C. Wagner finishes with these observations: "The ancients considered any builder a mason. They could see little difference between building a human character (THE SPIRITUAL), building a castle in the air (THE INTELLECTUAL), or building a structure of stone (THE MATERIAL). The "stone" that was lost during the building of the Temple was INTUITION, because there was no material evidence of its existence. And the temple of our soul (our true character) is never complete."

According to the "SECRET DOCTRINE" (H.P.B.), "When the Third (Lemurian) Race separated and fell into sin by breeding men-animals, these became ferocious, and men and they mutually destructive. Till then there was no sin, no life taken. After this separation, the Age "Satya Yuga" (Hyperborean) was at an end. The Eternal Spring became constant change and seasons succeeded. The cold forced men to build shelters and devise clothing. Then man appealed to the Superior Fathers. The Nirmanakaya of the Nagas, the wise Serpents and Dragons, came and the precursors of the Enlightened Buddhas. Divine Kings descended and taught men sciences and arts, for men cold live no longer in the first land Adi-Varsha, the Eden of the first Races, which had turned into a white frozen corpse." Thus, the Lemurian Golden Race, after the separation of sexes, developed the Monosyllable Language still in use among primitive yellow races. No longer could they communicate by thought transference or telepathy like the Hyperborean Paradisians. From the Monosyllable Speech was developed the Atlantian Agglutinative languages, that is, words formed by combining the original monosyllables to give complex meanings. The mystery tongue of the 5th Race was Sanskrit, which is described as inflectual speech, wherein word endings are varied to express different grammatical relations (verb tenses, and case, number, gender, comparison or declension of words) by similar sets of features. "At any rate," H. P. B. continues, "the Semitic languages are the bastard descendants of the first phonetic corruptions of the eldest children of the early Sanskrit...The (Jews) are a tribe descended from the Tchandalas of India, the outcasts, many of them ex-Brahmins who sought refuge in Chaldea, in Scinde, and Aria (Iran), and were truly born from their father A-Bram (No-

Brahmin) some 8000 B.C. The Arabs are the descendants of those Aryans who would not go into India at the time of the dispersion of nations, some of who remained on the borderland in Afghanistan and Kabul, and along the Oxus, while others penetrated into and invaded Arabia." Arabic writing is similar to Aramaic.

THE SACRED MYSTERIES OF ATLANTIS CONTINUED IN THE EGYPTIAN INITIATION

Iamblichus tells us that Plato was taught for 13 years by the School of the Egyptian Magi, giving his doctrines of Platonism which influenced Christian ideals; Thales, Pythagoras and Eudoxus also had passed phases of the Initiation, and Moses supposedly was raised by the Magi where he learned what later he used to formulate the Hebrew Teachings. The Trials of Initiation were extremely trying, and only the most capable, True Masters succeeded. However, no physical trials were submitted to the sons of the Magi, for education was their initiation, being in them because of their race. Sons of the Magi began studies when 15 years of age, and complete instruction lasted 21 years. Grades started from Theorist, Practicer, Philosopher, minor and major Adept, Master of the Temple, and by completion of studies they had become "Magus of the Rose-Cross". (Study page 47 with what follows:)

Each letter and corresponding number of the Sacred Language, which are called the 22 ARCANA, or keys, when contemplated or pronounced, expresses a reality of the Divine World, the Intellectual World and the Physical World. These Keys were later put to profane worldly use after the Egyptian priests began to use them to foretell the future together with astrology, and so that their Ancient Wisdom would be preserved for future generations, they were evolved into the 22 Tarot cards with symbolic pictures depicting trials and triumphs of life in mystical interpretation. Like the symbolic Chakras, Powers, Mandalas, etc. of Eastern Yoga and Buddhism, Western Hermetic Sciences require many volumes to describe, which we cannot enter into now. Beside the Egyptian Initiation by Iamblichus, we also have "The Tarot" by Mouni Sadhu which is an encyclopedic course of study on the subject, "The Book of Thoth" by Aleister Crowley, beside our study of

"The Emerald Tablets of Thoth" given in studies for my Doctorate in Metaphysics in 1949 at Lake Quilotoa. Dr. Anna Kingsford and Edward Maitland founded the Hermetic Society in Great Britain based on the Western Mysteries, while Mdme. H. P. Blavatsky and Col. H.S. Olcott founded the Theosophical Society in the U.S.A. to teach the Eastern Mysteries; both contributed much to systematically detail much of this information.

To inspire interest in the Minor Arcanum, and Major Arcanum of the symbolic pictures, I am going to describe the Privileges obtained by the Magi or Masters of the Arcana. First the Arcana numeral is given:

I. The Privilege to see God personally without dying, and speak freely with the Seven Planetary Spirits. Card is named Magus, Symbol = Will.
II. The Master Magician stands beyond the reach of all griefs and fears. Card named Door of Occult Sanctuary, Symbolizes Knowledge.
III. The Magus is Co-Ruler in Heaven and has hell at his service. The Card name is Isis, Venus Urania. Symbolizes Action.
IV. The Magus disposes of having health and life himself as well as that of others. Card named Cubic Stone. Symbolic of Accomplishment.
V. He cannot be surprised by destiny, is not tortured by misfortune and is not defeated by enemies. Card of Master of Arcana, has powers.
VI. The Adept knows the reason of Past, Present and Future. Card of Two Ways, or Temptation, which symbolizes trials of good and evil.
VII. The Magus knows the Mystery of Resuscitation of the dead and the Key to Immortality. Card of Chariot of Hermes or Osiris: Victory symbol.
VIII. The Adept has the Secret of the Philosopher's Stone. Card shows Themis, Goddess of Justice, or Law and Order. Equilibrium is symbol.
IX. Adept is in command of Universal Therapeutics. Card of Hermit and his Veiled Lamp (Esoteric Light). Symbolizes Prudence.
X. Adept reaches perpetual motion and quadrature of circle. Card shows Sphinx and Wheel of Fortune, symbolizing Fortune,

Good or Bad Life.

XI. Adept can turn not only metals, but also all refuse into gold. Card of the Conquered Lion. Symbolizes Force of Strength to Conquer.

XII. The Magus has dominion over all animals. Card of Sacrificed suspended from tree; symbolic of violent death.

XIII. The Adept possesses the Keys to all the Mysteries. Card with Figure of Death and Scythe, symbolic of Man's Transformation, Rebirth.

XIV. The Adept can speak scientifically and convincingly on all subjects without previous preparation. Solar Genius, symbol of Harmony.

XV. The Magus can judge a man from first look thru intuition. Card has Typhon (Egyptian Devil) which symbolizes unseen Fate.

XVI. The Master possesses the Arcanum of compulsion in relation to Nature. Card with Tower Destruction, symbolizing physical destruction.

XVII. The Master can foresee happenings dependent on Fate. Card with Star of the Magi symbolizes Hope, Intuition of coming Salvation.

XVIII. The Magus can bring solace to everyone in everything and give advice in all events of life. Twilight or dusk = Deception, illusion.

XIX. The Master will overcome all obstacles in life. Card with Shining Light or radiant sun, symbolizing earthly Happiness, Truth of Being.

XX. The Master overcomes all Love and Anger within. Card shows resurrection of dead, Judgment, symbolizing renewed trials of future life.

XXI. The Magus knows Secret to Riches, can possess them, but never is slave to them. Card shows a fool or dame Nature,- but the letter is a Crown of Golden Roses, symbolic of Supreme Mastery or Illumination.

XXII. The Master astonishes all laymen with ability to direct elements, cure the sick and resuscitate the dead. As noted in Arcanum XXI, the last two cards are confused. Iamblichus has "Crown of Roses" description, with naked woman, circle of wreath of greenery and man, eagle, bull and lion about it for Sichen or XXI, while Mouni Sadhu substitutes a Serpent under woman's foot for the wreath and places it with Arcanum XXII.

Thus, Iamblichus says XXII symbolizes a fool with bag of errors running into trouble, biting dog, crocodile and broken oblisque, while Mouni Sadhu says this is XXI. Obviously, the Hermetic traditions became confused with the systemizing of doctrines so that Oriental Wisdom could identify with Hermetic Mysteries, and account for origins of identity in Atlantian Mystery Schools.

When Missionaries of the Church of Rome first went to India and Tibet, they accused their holymen of having corrupted the Gospel of Christ, or the Devil had tricked them by identity of teachings of Sri Krishna and Buddha (without credit) to the True Savior. Hermetic symbols found in Ancient Egypt thousands of years before the Christian Era, described the basic doctrine of Christianity, often word for word. Actually the Gnosis of Logos and Gospels of Christ in reality were purloined from Hermetic Mysteries, which like the Hebrew interpretations were corrupted to conform with likings of the priest craft.

Let us quote works of Hermes Mercurius Trismegistus rendered into English by Dr. Anna Kingsford and Edward Maitland, (Virgin of the World) "Celestial Order reigns over terrestrial order...O sacred books of the Immortals, ye in whose pages my hand has recorded the remedies by which incorruptibility is conferred, remain forever beyond the reach of destruction and decay, invisible and concealed from all who frequent these regions, until the day shall come in which the ancient Heaven shall bring forth instruments worthy of you, whom the Creator calls souls...Heaven,- God manifest,- regulates all bodies. Heaven is full of God...Man then, Asclepius, is a great marvel; a creature worthy of respect and adoration. For amid this Divine Nature he moves as if he himself were a God. In joining himself to the Divinity, he connects himself by a bond of love to all other beings, and thereby feels himself necessary to the Universal Order. He contemplates heaven, and in this happy middle sphere in which he is placed, he loves all that is below him, he is beloved of all that is above. Perfection is obtained when the virtue of man preserves him from desire...If man be too much hindered by the weight of his body, he will be unable to penetrate into the true reason of things...For if the world is the work of God, he who by his care sustains and augments the earth's beauty, is the auxiliary of the Divine Will, employing his body and his daily labor in the service of the work produced by the hands of God... In the beginning were God and Hyle (matter or substance of the Universe). The things which constitute the Universe are not God, there-before their birth they were not in existence. The eternal God cannot and never could be born...Matter, being fecund in all attributes, is able also to engender evil..The Foundation of Knowledge is Supreme Goodness...Thought is a Light which illuminates the Intelligence as the sun illuminates the world. God's Will is Universal Goodness... Nature is born of His Divinity. God is the Father, the Universal Ruler. The Race of the Gods is formed of the purest part of Nature... The earth is an animated being...God is the Plentitude of Life...No thing in it is mortal. The sun is lasting as the Universe, and governs all living creatures, being the fount and distributor of all vitality. Intelligence, the tenacity of Memory, makes man the Lord of the earth."

Number of Atlantean Egyptian ARCANUM	English equivalent and hieratic name	Hebrew letter and name.	Aramaic Letters in Jacobite & Estrangelo characters and names.
I 1 (A)	Alohim	Aleph	'Olaph
II 2 (B)	Beinthin	Beth	Beth
III 3 (G)	Gomer	Gimel	Gomal
IV 4 (D)	Denain	Deleth	Dolath
V 5 (E)	Eni	He	He
VI 6 (U. V.)	Ur	Vau	Waw
VII 7 (Z)	Zain	Zain	Zain
VIII 8 (H)	Helitha	Cheth	Heth
IX 9 (T. H.)	Thela	Teth	Teth
X 10 (I. J. X.)	Iojni	Yod	Yudh
XI 20 (C. K.)	Caithe	Caph	Coph
XII 30 (L.)	Luzain	Lamed	Lomadh
XIII 40 (M.)	Mataloth	Mem	Mim
XIV 50 (N)	Nain	Nun	Nun
XV 60 (X)	Xiron	Samech	Semcath
XVI 70 (O)	Olelath	Ayin	e
XVII 80 (F. P.)	Pilon	Pe Phe	Pe
XVIII 90 (T. S.)	Tsaid	Tzaddi	Sodhe
XIX 100 (K. Q.)	Quitolath	Koph	Kuph
XX 200 (R)	Rasith	Resh	Rish
XXI 300 (SH)	Sichen	Shim	Shin
XXII 400 (T)	Thoth	Than (Tav)	Tau

(Publisher's Note: To see the script of Aramaic (in Jacobite and Estrangelo characters) and Hebrew letters, consult the following web addresses: http://www.omniglot.com/writing/syriac.htm and http://www.omniglot.com/writing/hebrew.htm)

After the Roman Numeral of each Arcanum we give the number it represents in Egyptian Sacred language, as well as in Hebrew and The Aramaic equivalents. The last letter (Thoth-Tav-Tau) is also the number 400. One can find letter designs of Hebrew characters, beside Arabic, Greek, etc. in Webster's Dictionaries to compare Alphabets. The Hebrew language is a double tongue; (1) The Ancient Language with consonants only as found in the Septuagint Bible, and (2) the more Modern Innovation with added vowel-points, such as found in the Masoretic Bible text, which came into use one thousand years after the Septuagint. Thus, as witnessed above, the dialect theologians study today as "Hebrew" is not the original Ancient Tongue, but a composite language adopted in the Babylonian Captivity of the Jews, or Aramaic-Egyptian corruption, enabling a falsification of Ancient Essene Documents.

Number of Atlantean Egyptian ARCANUM	English equivalent and hieratic name.		Hebrew letter and name.		Aramaic Letters in Jacobite and Estrangelo characters and names.	
I 1	N	(A) Alohim	N	Aleph		'Olaph
II 2		(B) Beinthin		Beth		Bēth
III 3		(G) Gomer		Gimel		Gomal
IV 4		(D) Denain		Deleth		Dolath
V 5		(E) Eni		He		Hē
VI 6		(U. V.) Ur		Vau		Waw
VII 7	M	(Z) Zain		Zain		Zain
VIII 8		(H) Helitha		Cheth		Hēth
IX 9		(T. H.) Thela		Teth		Tēth
X 10		(I. J. X.) Iojni		Yod		Yudh
XI 20		(C. K.) Caitha		Caph		Coph
XII 30		(L.) Luzein		Lamed		Lomadh
XIII 40		(M.) Mateloth		Mem		Mim
XIV 50		(N) Nain		Nun		Nūn
XV 60		(X) Xiron		Samech		Semoath
XVI 70		(O) Olelath		Ayin		'ē
XVII 80		(F. P.) Pilon		Pe Phe		Pē
XVIII 90		(T. S.) Tsaid		Tzaddi		Sodhē
XIX 100		(K. Q.)Quitolath		Koph		Kuph
XX 200		(R) Raeith		Resh		Rish
XXI 300		(SH) Sichen		Shim		Shin
XXII 400	T	(T) Thoth		Than (Tav)		Tau

After the Roman Numeral of each Arcenum we give the number it represents in Egyptian Sacred language, as well as in Hebrew and The Aramaic equivalents. The last letter (Thoth-Tav-Tau) is also the number 400. One can find better letter designs of Hebrew characters, beside Arabic, Greek, etc. in Webster's Dictionaries to compare Alphabets. The Hebrew language is a double tongue; (1) The Ancient Language with consonants only as found in Septuagint Bible, and (2) the more Modern Innovation with added vowel-points, such as found in the Masoretic Bible text, which came into use one thousand years after the Septuagint. Thus, as witnessed above, the dialect theologians study today as "Hebrew" is not the original Ancient Tongue, but a composite language adopted in the Babylonian Captivity of the Jews, or Aramaic-Egyptian corruption, enabling a falsification of Ancient Essene Documents.

MYSTICAL ANTHROPOLOGY CONSTITUTED AS A FUNDAMENTAL VITALOGICAL SCIENCE
by Dr. Johnny Lovewisdom

CHAPTER VI: ANCIENT LEGENDARY HISTORY OF THE KRISHNA CULT, AND ARYAN CULTURAL BENEFICENCE FROM BUDDHA

Would you take a curly headed black man as your Savior, if he were so powerful he could murder the king of your nation, besides thousands of his noble subjects, besides being a leading thug in a gang of assassins who not only joined him in killing his close relatives and teachers, but went about terrorizing and butchering men and animals, and led a white man to fight a war against his own relatives after he threw down his weapons declaring he would rather be killed than kill his own people? Further-more, not only must one worship such a notorious homicide, but also one is taught to accept his gospel as the very Truth, in which doctrine one can do all kinds of actions, good or evil, as long as one acts indifferently without attachment to the actions or the victims, deriving neither pleasure nor guilt in doing them. As witnessed, this is the belief of all who adopt crime and seeking unearned pleasures as their career.

What's more, this has become the Saint or Savior of millions who become disciples of Yoga and Krishna cults of Hinduism. Like the marching music of military bands, the Gospel of Krishna, with bestial sounds, harmony and melody, soon overcome the most peaceful and poor who have nothing to gain by it, but yet take up arms and go to war. Yet, in modern times people think of India, Yoga, and Hindu Religion as a morally vegetarian nation dedicated to peace. Actually if Hindu chronology be trustworthy in dating Krishna's time to 5000 B.C., it is only the latter half or over 2000 years that India has become a pacifist-vegetarian population due to the Gospel of Gautama Buddha, which was made the Law of India by king Asoka, and re-awakened by many teachers in a modern era of Mahatma Gandhi. Buddha was diametrically opposite to such teachings in Krishna's doctrine, preaching against spreading of his doctrine by bloodshed, as well as against eating slaughtered food or any justification for destroying life.

Much like Moses and the prophets of Temple Slaughterhouses of Jewish subjects, Krishna and his followers embellished the doctrine of the beef-fed Aryans in the conquest and subjection of India. "Apparently it was early in the second millennium B.C. that the main bulwarks of ancient culture were broken. Then it was the Aryans invaded the Punjab from the north, the Hyksos gained control in Egypt, the shepherds spread over Palestine and the barbarians sacked Crete." Thus, wrote Henry Bailey Stevens in THE RECOVERY OF CULTURE, adding, "The Aryans who swept into Punjab literally recognized the association of war with more cows, and the austere beauty of the hymns of the Rig-Veda is contaminated with the rituals of flesh sacrifice...Man had gone into business as a Butcher...The truth is that animal husbandry and war are institutions in which man shows himself most proficient. He has been the butcher and the soldier, and when the Blood Culture took control of Religion, the priestess was shoved aside." In contrast to the climax of the "Lord's Song (Bhagavad Gita) of the battle and cattle crazed Aryans in the Mahabharata, Stevens elucidates: "The Buddha concentrated his attention particularly upon the BUD stage, for this is the period of awakening and is the period of the naturally pure. The essential achievement of the BODH tree (under which he achieved Illumination) is that it has succeeded in keeping this pristine quality thru-out the season, ward off the very entrance of evil, which would otherwise gain foothold and gradually spread its infection until it rots the fruit. So the mind must resist the wrong thoughts which turn into wrong acts, even preventing their entrance, if it is to learn the lesson of the Tree of Wisdom (Bodh)...Since the shedding of blood is man's greatest evil, it was his first vow not to partake in killing in any way. He refused to eat flesh. And going forth he proclaimed himself free from the chains of the world."

Much like the Gospel of Christ, Christna is sought by king Kansa, the Hindu Herodes in the slaughter of infants, which Christian missionaries claimed was copied from the life of Jesus, when the reverse was really the truth since Krishna preceded Christ by millenniums. Similar to the legend of how the high priest, Zachary, fathered the birth of John the Baptist, whose mother was Mary (since his own wife was barren) and then betrothed the Temple Virgin Mary to Joseph, in whose house he was known as Jesus or Healer rather than John. This enabled the mystical allegories for students reading the Gnostic Essene Scripture, but the Esoteric disciple solves the enigma, he understands why Herod identified Jesus as John the Baptist, as did others recognizing the human aspects, altho John's teaching was that with the Baptism of the Jordan he was Anointed or "Christ-born" with the Holy Spirit of Truth to become the "Healer-Savior" (Jesus) that the world awaited. Christ was of Holy Spirit, not born of flesh, so flesh-birth could not identify Him. Since Herod himself was tutored in an Essene School, a classmate of John, he could only fake the incident of the beheading of John, requested by Salome, being far from his own desire and honor among Essenes. Later, the mock crucifixion was enacted in a similar manner by Pilate to calm the Jews, actually crucifying Simon, the Zealot Jewish warrior.

Likewise, Kansa the King was Krishna's uncle, who not only imprisoned Krishna's father and mother but sought to kill Krishna, whose parents had saved him by hiding him to be raised by cowherds. Many who take up the study of the Gita and Yoga, sense that in spite of it being a Spiritual allegory, the historical record of man's inhumanity to man and beast, smothers and destroys any such classification. The details identify a passionate playboy, the black darling who grows up to be a heroic warrior and ruler at the best. In Swami Sivananda's version of "Srimad Bhagavad Gita", in Krishna's defense he elucidated on accusations such as "He is not an Avatar: he is a passionate cowherd who played with the Gopis (milkmaids). (Swami) "Was he not a boy of 7? Can there be a tinge of passion in him"? Thus, the Ras Lila, or this type of God-Play or allegory is presented as the exemplification of devotion to Krishna, loving him as the Gopis did. Not only did Krishna and his brother Balarama play with the Gopis, but they went around killing men and animals who they claimed were devils seeking to destroy them. The Swami then tells of, "Some people catch fish in the Ganges for satisfying their palate and quote the Gita to support their action: 'Weapons cleave him not, nor fire burneth him.' Devils quote Scriptures!"

After reading parts of the Old Testament of the Bible, justifying all kinds of violations to the Ten Commandments by the Jewish men of God, because the Jews were God's Chosen people and Jehova told them to contradict his word, I could only identify the Bhagavad Gita with the same double-talk-concerning those who were labeled "Nobel" and in Union with God or "Yoga", when I studied them in Quilotoa 30 years ago. Why study doctrines starting out and basing their doctrines on such obvious hypocrisy. Such may be fine for people now lost in their own passions, who can only be reached by discussing their own interests, which such scriptures seek to lead one away from or mitigate, but in turn many of us in the New Age have Intuitively evaded getting involved in a life of pleasure-seeking at an early age. It is not that I do not understand the Lilas used to capture the mind, but we already have hundreds of millions of followers of the Hebrew-Christian Bible and the Hindu Gita, who have adapted Scriptures to justify their pleasure seeking lives as men kill, rape, steal, deceive, etc. In turn, we seek the regeneration of mankind, the restoration of Paradise and Truth.

For instance, after Krishna kills his uncle Kansa, he fights 17 battles, killing all the overwhelming enemy troops and not losing any of his select army. In the 17th battle we are told that all the infantry fall in the battlefield, among the enemy, with their bodies cut to pieces. Many thousands of elephants and horses likewise were butchered. The flow of blood came by waves, and in this river the arms, cut from shoulders appeared like snakes, the heads were like turtles and the dead bodies of elephants were like islands, the horses appearing like sharks therein. All this flow in the river of blood filled with all these things had been arranged by the Supreme Will. The hands and feet of the soldiers appeared like seaweed! Such a picture of Lord Krishna's pastimes making war does indeed picture/sound like the news we hear every day from the world's battlefronts, along with righteous justifications presented by either side. There is only modern sophistication in weapons and recording to entertain the world. We have lived for several Millenniums in what is called Kali Yuga, the Age that Krishna founded, with the passionate wars and sensuality he exemplified without learning how to put an end to passionate desires, in willful abstinence from their causes as the Lovewisdom Message teaches.

Yet Krishna said, "He who puts his trust in me, never perishes.

Even he with the worst Karma who forever meditates on me, quickly loses the effects of his past bad actions." Just to meditate on Krishna killing hundreds of thousands of men and beasts alike, and his playing with the Gopi girls to become India's Savior, instead of starting Indian history with spiritual beginnings, shows the Hindus only adopted the Savage black man's ideal of Decca (India) and Africa's jungles inherited from post-Lemurian peoples "black with sin". How the fair white Aryans got involved into this, intermarrying with these black natives has been an ethnological mystery.

The changing of the earth's axis, brought the glacial ages of ice and snow to the former northern Polar Paradise of Hyperborea. By accident, humans were not destroyed by sudden freezing when the cataclysm precipitated. In protected areas man noticed animals survive if protected, and thus not only built warm homes or caves for himself as well as for cattle, beside provide them with dried fodder or hay for the winter months. In return, cows furnished milk which kept well if soured, and thus made wild greens, starchy vegetables, etc. more palatable, and the male animals served for transportation and clearing large fields to raise the yearly food supply in the few warm summer months. In the extreme cold man no longer selected juicy fruits and herbs as instinctively done where one perspires freely, but instead he sought the dry concentrated ones that did not perish rapidly, seeking to eat large tree seeds from the forest, but this led to selecting the seeds, rather than the juicy pulp of fruits. The stone of the peach was cultivated for the seed which became the almond, and the walnut apple not being edible was esteemed for the protected seed or walnut. The juicy fruits that did continue in esteem were more to please the taste rather than being selected for all around food value, as well as being selected for keeping qualities in the case of apples, and for drying in the case of prunes, raisins, figs. etc.

In Lemurian times before 70,000 B.C., some of the fair and tall mighty Hyperboreans wandered south to become the South Sea Islanders who remained more on the fruit, vegetable food and roots. Later migrations in Atlantian times after the glacial ages brought in the evils of cooking foods, eating slaughtered flesh and seed substances such as nuts, grains, legumes and their products.

After nut and seed cakes, grains were found palatable in bread and soups, after men took to the use of fire to warm their homes and make food warm in the winter's cold. Radioactive Carbon tests show that grains were developed from wild grasses, by selecting the largest seeds, in the Atlantian Age, during the 10,000 to 15,000 B.C. period. However, hunting wild animals for food pictured in cave pictures date back 20,000 years ago in Siberia and Southern Europe. Geo. G. MacCurdy remarked, "The monotony of the food supply prior to the Neolithic period (about 20,000 years ago) seems appalling: In Switzerland during the last inter-glacial epoch it was 90% cave bear; in Moravia it was 90% mammoth; in Denmark it was 90% shellfish." In turn, Stevens observed, "There is no evidence of warfare between the hunters and the horticulturists. Economically their interests were not at variance. It was easier to grow crops successfully when the wild creatures were kept in abeyance. So the hunters and fishers, tho they were at war with most of the animal kingdom, were at peace with their fellow men, even as are the Eskimos today..." Eskimos, whose very name means "Eaters of Raw Flesh", today number 35,000, half the capacity of the Yale Bowl, Stevens adds. From this we may conclude that meat eating due to lack of other food did not develop the perverse warlike and sensuous human traits, but this came about with the use of seeds, predominantly grains.

It is evident from the writings of modern teachers of Yoga and Hindu philosophy, that a great deal is being hidden as to the origins of Aryan Culture claimed for India, as to diet and the rapid degeneration of the original white race that invaded India. It is much like the cover-up invented by the early supporters of the Roman Church: Instead of allowing the allegorical teachings of the Gospel to remain in their original form, in which the novice must develop his intuitive and truth finding faculties thru the Mysteries, the embryonic "Holy Roman Empire" founders prevented the clear vision that John the Baptist was the vehicle manifesting Jesus Christ. Dawn and Baptism are expressed by the same word in Aramaic, because the Dawn is the pouring forth of Light or Illumination, and Baptism is the pouring forth of the Living Water in the Spirit of Christ in Gospel Allegories. John means the God-Endowed, who by Baptism of the Christ Spirit became the Dawn of Enlightenment or the pouring forth of Illumination, verily a

Buddha. thus, after his 29th year he was known as the Healer (Jesus) Anointed by the Spirit of Truth (Christ). Rome's defenders thus used only meanings to words that would completely change the original N.T. story: Today Christians believe the Savior took wheat bread when he blessed his own St. John's Bread or carob, and when he speaks of meat they think of the Pasqual Lamb when he really spoke of food. The mother of Constantine, Helen, lamented that there was no sign of any Jesus Christ having existed in all of the Jewry in Palestine, and there upon made it her mission to designate where the Savior was born, was crucified and buried, beside other events described in the Gospel Allegories. There had been no town of Nazareth prior to this, so once designated, it grew into being, and twin towns like Bethany on the Jordan and Bethany near Jerusalem became necessary to divide the evidence that John was Jesus into appearing as two individuals. The symbol of the Cross is not Christian, used thru-out pagan and heathen history. Its earliest known form of a human figure on a cross is a crucifix presented by Pope Gregory (590-604) to Queen Theodolinde of Lomabardy. No crucifix has been found in the Catacombs, altho the swastika cross was found there with "Vitalis Vitalia" (Life of life) inscription. The earliest representation of Christ Jesus was a figure of a Lamb, but by the 6th synod of Constantinople it was ordained that instead of the Lamb, a figure of a man fastened to a cross should be represented as is found in pagan religions. The earliest artists of the crucifixion represent the Savior as young, beardless, alive, elated and with no sign of bodily suffering. The cross in Egypt was used as a protecting talisman and symbol of SAVING POWER. Typhon, their Satan, is represented as bound and chained by the cross. Thus, in the Initiation into the Mysteries the "Crucifix" is placed horizontally on the ground where the Neophyte was bound on this couch of Trial to be tested as to his Faith, Purity of heart and Mastery.

Krishna is also represented as a black man crucified with arrows instead of nails. Alex. Von Humboldt speaks of the Mexicans celebrating a feast with a black man pierced with arrows in crucifixion, and other authors verify that Quetzalcoatl is honored in this crucifixion symbol, which greatly amazed the Spanish monks who first arrived with the Conquistadores. It proved the Mystery of Crucifixion Initiation circled the earth among mystical societies, and was not in any way original with Christianity. An early Christian Father, M. Felix, in argument with an infidel, affirmed: "I must tell you, that we neither adore crosses nor desire them,...You it is, ye Pagans who are the most likely to adore wooden crosses... crosses gilt and beautiful. Your victorious trophies not only represent a simple cross, but a cross with a man upon it." Tertullian likewise claimed of Pagans: "The origin of your gods is derived from figures molded on a cross." This doctrine of the crucifixion made Jesus Savior of Rome, embellished with interpolations in the N. T. such as, "I came not to bring Peace, but the sword", and the doctrine that he went forth fearlessly unto death to defend official Roman Church doctrine, giving his life for the sins of mankind. This has been used by all Christian nations ever since, Faith in supporting crusades, the Inquisition, the Conquistadores, World Wars, Korea, Vietnam, and all wars by Christian nations.

Going into death by taking up the sword to defend Christ, reveals origins in the doctrine of Christna. "In this world there is a two-fold path, as I before said, O sinless one: that of Yoga by knowledge, of the Sankhyas; and that of Yoga by action, of the Yogis. Thus, Sri Krishna begins to teach his path, when Arjuna throws down his weapons refusing to kill kinsmen in battle. "Therefore, FIGHT, O Bharata. he who regardeth this as a slayer, and he who thinketh he is slain, both of them are ignorant. He slayeth not, nor is he slain. He is not born, nor doth he die; nor having been, ceaseth he anymore to be. Unborn, Perpetual, Eternal and Ancient, he is not slain when the body is slaughtered. Who knoweth Him, Indestructible, Perpetual, Unborn, Undiminishing, how can that man slay, O Partha, or cause to be slain?"

Thus, Krishna develops the doctrine of the Higher Self, the Immortal God-Being within. The pious path of religion requires absolute decision. "Didache" (Doctrine of the 12 Apostles) starts out quoting the tenets of the early Carmelite Prophets: "Behold, I set before you two ways, the way of Life and the way of Death!" Either one decides to go all the way in the renunciation of one's desires to Behold God in the Way of Life, or if one falters, still seeking any worldly joys or pleasures, one must go the Way of

Death. As long as there is attachment, as long as desires prevail, men work in perpetual destruction of themselves. Only when all become Renunciates will men live in Paradise without war, but otherwise Krishna and Christ require war or Conscription. Ecology has proven that Life and Death are 2 paths. What Krishna seeks to do is assure the young soldier that he need not heed the intuitive conscience which protests against killing and sin in the purer and more sensitive Aryan, since putting trust in this black magician, one's karma is cancelled,- as wild imaginative tales should indicate. Otherwise, he must evade contact with civilized life enjoying the fulfillment of desires or pleasure,- living as a hermit hidden in the forest. A similar policy of Conscription exists still.

In Edward Schure's "Ancient Mysteries of the East" we read: "From the conquest of India by the Aryans, emerged one of the most glorious civilizations the earth has ever known. The first of Messiahs, the eldest of the sons of God was Krishna. By reconciling the two warring groups, the white race and the black race, the solar cult and the lunar cult, this divine being was the true creator of the national religion of India. The battle between the sons of the sun and the sons of the moon, between the Pandavas and the Kuravas is the theme of the great Hindu Epic, the Mahabharata... In the middle of the gigantic epic, the Kuravas, the lunar kings become the conquerors. The Pandavas, the noble children of the sun, are dethroned and banished as exiles, they hide in forests seeking refuge among the anchorites, wearing clothing made of bark and leaning on hermit's sticks." Then Schure describes how Devaka, Krishna's mother, went to live with Vashichta, leader of the anchorites, to escape death from her brother, King Kansa, who is told she will bear the master of the world. "Vashichta had the venerable appearance of a god. For 60 years he had eaten only wild fruit. His hair and beard were white as the summit of the Himavat, his skin transparent..." It was there that Devaka gave birth to Krishna, living from the wild fruits of the forest. The ideal is beautiful, but if Krishna's mother lived on <u>wild fruit</u>, it does not account for Krishna being a black baby, as we have shown in earlier chapters. Young Krishna adored his mother, wrestled with panthers and became a strong youth. When 15 years of age his mother disappeared, and Krishna left the hermitage for a life making love to the Gopi girls, and fighting battles till he became ruler of the country he lived in. In Schure's version, Kansa shoots an arrow at Krishna which misses and kills Vashichta. In the end, yellow and black soldiers tie Krishna to a tree and crucify him with arrows.

In contrast, "The Bhagavad Gita" by W. Q. Judge claims that the Kurus were the Aryans from the region between the polar sea and the snowy Himalayas, altho agreeing that the Pandavas represent the spiritual side of man, while the Kurus represent the lower nature. Mdme Blavatsky, Doane, and various other well versed writers, show that the word "Krishna" means a black man, explaining why he was so named, beside having curly or kinky hair, and having 4 symbolic arms, a flute, and his first name was Govinda, meaning cowherd. This black cowherd could hardly have been only 7 years old, when he went to play with the Gopis, Schure's estimate of 15 years being apt for starting the romantic tough life of a cowboy, herding cattle and enjoying a love life with the milkmaids, later followed by his warrior ventures.

Next, referring to A.C. Bhaktivedanta Swami Pradhupada's book on "KRISHNA", together with defending Krishna's life as a warrior killing hundreds of thousands of men and animals, he lauds the black man by these qualifications as proof of his supernatural personification of God and the God-Self. Yet, one does not need to study Krishna to know and realize the God-Self within one, since this teaching prevails in religions before Krishna's time as well as those after him. To the contrary, we might sow the impurity of this reprobate's teachings, not only of the lowest morality in his actions, but all admit that this magician was a black one. The dark skin, dark eyes and black kinky hair all denote that his diet was cadaverous, and any linking of this black cowherd with Aryan forefathers can only be in their pork and beef-eating habits. Swami Pradhupada, who claims lineage from Krishna, was a black who speaks with ardor for Race-mixing, for a Lila of Maya (play of delusion) for solving racial problems. As we described in earlier chapters on racial traits, cooked grains and flesh make men dark.

In 1967, Bruno H. Schubert wrote in "SURVIVAL OF MANKIND", "The majority of India's people live in poor health.

About 150 million live on one meal a day consisting of bread and soup. The average life-span of these people is 27 years, the lowest of any race. They are degenerated physically and mentally for lack of trace minerals which are missing in their starchy diet of rice and wheat. India has the richest and deepest top soil on earth. Starvation is going on in midst of plenty." People who speak of Yogis and sages of India are deluded by glib tongues of those who have not solved when the simplest problems of life, and dietetically degenerate themselves into utter misery. It was fortunate that the Hindus held the cow sacred, compared to other animals. Schubert continues, "Cattle eat the tops and part of the stems of the grass and herbs in the pastures, but leave the lower part intact for further growth of the plant. BUT SHEEP, GOATS AND CAMELS EAT EVERYTHING THAT EXTENDS FROM THE GROUND. THEY EVEN PULL OUT THE ROOTS." This overgrazing by goats, sheep and camels made the Gobi, Sahara, Arabia, etc. into barren desert. "Dung, dung, dung! It is everywhere in India. One can see a worker following cattle with brush and basket to collect the manure as it drops on the street. The worker then pat-a-cakes the dung into fuel or building blocks for his house. In miserable dank shawl or slums of India's biggest cities, the workers cook, give birth and die between heaps of rotting garbage and stinking pools of sewage and dung. Everywhere you see it. Everywhere BUT IN THE FIELDS where its valuable life-giving properties could do the most good." Schubert's evaluation of Hindu culture goes far in explaining why I was never inspired to go study with Hindu Yogis. The life- style of Yogis illustrate ignorance, degeneration and misery of pious black magic of ancient stubborn origin.

Swami Pradhupada does concede that the Brahmanas, or the highest caste, are not allowed to have an illicit sex life, eat meat, gamble, use drugs and alcohol, etc., but this reflects REFORMS THAT THE BUDDHA PRESCRIBED, and the father and grandson kings Asoka made law thru-out the vast territory of India, extending from Ceylon to Siberia, which was extended west to include Egypt and Syria-Palestine. King Asoka II or Asoka of the Good Law, Dharmasoka, was the grandson. Edicts carved in stone forbade the killing of humans and animals thru-out all these lands.

"KRSHNA" (Study of the 18th Song of Srimad Bhagavatam of Vyasadeva), of Pradpata shows that the diversions of Krishna include enjoyment in the hunt of wild animals of the forest, which he claims necessary to keep peace in the forest and to obtain flesh for sacrificial fires in their temples. This enabled the development and maintenance of the callous and ruthless nature of Krishna's Warrior Caste, the Ksatriyas, so the they could kill their enemies without mercy supposedly to preserve peace in human society. These warriors had to be able to kill their own brethren without wavering! It is evident by such low moral standards that these primitive black men of the Lemurian race that populated India, the East-Indies and Africa, after the sinking of the continent, were not far removed from carnivorous beasts. When Krishna kidnapped his first wife, Rukmini, he became involved in fighting with all his opponents, but so great a warrior was he, it is said, that millions of heads decorated with helmets and earrings covered the battlefield beside camels, elephants, horses, weapons, chariots and scores of other things they had, and all this to get but one of his 16,108 wives! When he went to search for a stolen magic jewel, "Syamantaka", he had to fight the most powerful being alive, Jambavan, the King of the Gorillas, who nevertheless was a devotee of Krishna otherwise, which was entered into for sport, but when the gorilla-man recognized the Supreme God-Self as his opponent, he weakened and gave in. Jambavan had given the jewel to his child, from whom Krishna was about to snatch it before the fight, but now the gorilla gave it back to Krishna, along with his daughter in marriage. Having ten children with a young female gorilla may achieve more than recent black kings in fighting, Ali Khan, "the greatest", and Joe Louis, but this captivated the primitive minds of ancient India. Such depravity of sexual degenerates triumphing by brute bestial force prevails in the law of the jungles of India as well as worldwide, from Krishna to the recent death throes of Kali Yuga. To claim this is the God-Self forbids nothing!

Then, to top this off, there is the claim that Krishna is self-satisfied, that he doesn't need a wife, and has no love for women, which again speaks for his depraved concepts of diversion in his favorite pastimes; but worse characterized his women as prostitutes who are only sought out for sex, and not out of love.

Being omni-present as the god-self, he was self-satisfied by omni-present other-halves of the female gender. Sometimes Krishna is claimed to have hundreds of thousands of wives, while usually he is limited to 16,108 which is said to be permitted by Vedic culture which allowed Polygamy. Next, considering that Krishna is said to have lived over a hundred years (125), and that he had 10 children with each of his 16,108 wives, this means that every day of his life he made at least four women pregnant, in what is lauded as the exemplification of how a good chaste family man should live! To get more women pregnant each day would have taxed even a Yogi's imagination, as well as even a gorilla's potential sexually. As the text reads,- "Krishna was so chaste he never had any sexual relations with women except to bear offspring." He was too busy conceiving to waste time otherwise.

Each wife had hundreds of thousands of servants, who took care of the thousands of Krishna's children and wives, while the wives spent their time massaging, bathing, perfuming, fanning, feeding and entertaining the beautiful black "God-Self". Just the looks of these women, who reddened their lips by chewing red betel nut, which when they spit it out appeared like human blood, must have made men blind as to whether they were engaged in bloodshed on the battlefield, or bringing more children into the dark starving masses of India's population which Krishna's seed engendered for thousands of years of sad misery among her depraved multitudes. The greatest possible degeneracy and shortest lifespan of these starving masses thus is visible in their traditional slavery providing such luxurious living for the opulent rulers, their many wives, and even more numbers of growing children, all of whom require even greater number of servants, and all of which were supported in food, clothing, palaces, jewelry, and every desirous whim, by a starving work-force of millions of laborers. This past Kali Yuga has been an era glorifying the "Struggle of the Fittest", the Law of the Jungle which reduced man down to degradation inferior to his recently adopted "cousin", the gorilla, who with the other anthropoids refuse to bear offspring in such oppressed environs. Peace will only come with the refusal to kill and oppress, meaning a refusal to fight, to be angry and to hate. Plenty will come by refusing to create heirs to divide the earth's wealth into scantier and more miserable allotted portions; and provide instead, greater rations of food, clothing, shelter, land and unpolluted environs for those now living. Saviors, as defined in Aramaic, should be Healers of mankind, and Krishna-like warriors who pretend to prove their Godhood by mass-slaughter, number of wives, offspring and servants beside a great waste of wealth for pleasures, requiring hundreds of millions of slaves living in starved misery to provide such opulence, must go and not be exalted and glorified as the Masters of Godhood and Yoga.

Instead of finding Godhood within themselves, for these millions of simple cowherd minds of Black "Aryans" of over 2000 B.C. to recent time, Krishna was supposed to be an incarnation of Vishnu, 2nd Person of Hindu Trinity, Sustainer or Savior of his people. Yet, since he could give birth thru thousands of wives he made himself equal to the First Person of this Trinity, or Brahman, the Creator, and moreover since he could kill or destroy his mightiest offenders, he is pretended to be the 3rd Person of the Trinity, or Siva, the Destroyer and Transformer. By Might, or by Hoary Legend he personified or incarnated all, Brahma, Vishnu and Siva, his followers claimed in substance making him their Highest Almighty God. Today, many pass all this off as exemplification of the God-Self potentially within us all, but the exaggerations of the Krishna Legend certainly must have been the tool that was exploited to recruit millions upon millions of men into the armies of India's rulers, and into the slaving masses of their laborers. Casting birthmarks of caste and hypnotic religious subjection upon so many of earth's inhabitants also followed organized Christianity to shower wealth on Popes and Rulers of the Holy Roman Empire, whose responsibility is easily shrugged off as too "sacred" to be discussed. Flesh and blood can never become Infinite Almighty Omniscience we may describe as God. The Spirit may manifest in all things as the Hidden Truth, but we cannot name or claim any mortal person, whether Krishna, Christ Jesus, John, Gautama, Hermes as God by physical attributes or actions altho idolatrous religions want to confuse past legend with such, rather than allow God to manifest his Spirit within us here and now present. Mixing love and devotion with war gave fighting nations.

The Indus cities of Northwest India were established by 2400 B.C. we are informed by Carbon 14 analysis, in data from "The First Great Civilizations". (J. Hawkes). Findings in excavations at Mohenjo-daro and Harappa include a prototype of Siva in his aspect as Pasupati, Lord of Beasts. The god has 3 faces, a huge horned headdress and is seated with soles of feet together and knees outspread. Each house had a bathroom and privy, earthenware sewage pipes took away waste, and the unusual cleanliness of these early Indus citizens speaks well of their culture. For food they did raise barley, wheat, wheat, rice, peas, sesame, melons and dates, along with pigs, sheep, humped cattle or zebu. However, in the 3rd century B.C. it is claimed that an Indo-European people invaded the Indus Valley destroying everything on their way southward into central and south India.

In "THE BIBLE OF THE WORLD" (Ballou), the Vedas are dated as appearing 1000 B.C, the Upanishads are dated about the 7th or 8th century B.C., while the "Bhagavad Gita" was inserted in the Mahabharata as late as the 2nd or 3rd century A.D. Thus, one may dispute from either side, as to whether the Legends of Krishna borrowed from Christian origins, or if legends of Christ borrowed from Krishna, and if they had not mutually borrowed from somewhere else.

The teachings of Siddhartha Gautama Buddha appealed more to reason and man's spiritual aspirations in Peace, Truth and Wisdom. To give a summary of Buddhist origins, again I will cite data from "The Bible of the World". Gautama was born 560 B.C., a prince of the Sakya clan, who at the age of 29 forsook status of becoming a future ruler, leaving wife, child and palatial home to retire to the forest. There, after 7 years of meditation, seated under a tree, he came to those conclusions which as a founder and leader of an order of followers called Bhikkhus, he expounded to them and the world. It was because of this that he received the title of BUDDHA, which is not a name but an appellation meaning the "Enlightened" or "the Knowing One" (Gnostic). Gautama's doctrine is a composite of ethical rules and philosophical approach to the riddle of human existence and the problem of human misery. It does not recognize a Hindu caste system, denies or evades the conclusion of Hinduism concerning an intangible supreme being with its coterie of personified powers in the form of gods, and

RENUNCIATION OF ALL DESIRES as the way to achieve the cessation of misery, thru NIRVANA (defined variously as "passionless Peace", extinction of the individual consciousness thru merging with the Infinite, the Void, etc.). It places the blame for an undesired continuation of existence thru rebirth (reincarnation) not upon fate, God, the devil, or heredity, but on the individual thru the law of Karma (continuing effect of one's past deeds). The world about us is all delusion, the ego is not genuine, not a true reality, but only a degrading composite of temporary delusions.

Yet with all of the socially negative emphasis upon forsaking the world, withdrawal from society, celibacy, and a materially non-productive life, there is in Buddha's teaching the highest degree of human COMPASSION that may be found in other religious doctrines.

The statement, "Ahimsa paramo Dharma", or Non-injury is the Highest Law, and Truth also, implies that the moral factor is the backbone of all law. The rights of man are grounded in a law superior to the laws of the State. To the Buddhist killing human beings, in war or any other reason is the worst offence or crime, and in this regard of Sanctity of Life, animals are no exception. To eat meat when all kinds of vegetable food is available, denies that one understands the Dharma. The positive aspect of Ahimsa is Maitri, or brotherly love, feeling of kinship and friendliness. In the Surangama Sutra, Buddha unmistakably teaches: "You of this great Assembly ought to appreciate that those human beings who kill sentient beings and eat their flesh, are not true disciples of Buddha. Ananda, next to teaching the people of the last age to put away all sexual lust, you must teach them to put an end to all killing and brutal cruelty. Besides being kind to animate life, one should refrain from injuring insects or even herbs."

The PANCHA SILA or Five Rules of Conduct are the Vows of a Buddhist. They are: (1) I take upon myself the vow of abstaining from causing hurt to living beings; (2) I take upon myself the vow of abstaining from taking that which is not given by its owner; (3) I take upon myself the vow of abstaining from sexual intercourse or giving birth to other sentient beings who suffer as all those who are born into this world; (4) I take upon myself the vow to abstain from lying, falsehood or deception, (5) I take upon myself the vow of abstaining from alcoholic drinks, drugs and other things that confuse and distort thinking and the mind.

Killing of animals for food, sport and sacrifice in Hindu temples was enjoined on the people of India, as well as worldwide practically, by religion and state laws. The Buddhist Asoka decreed laws and appealed to mankind to outlaw the cruel practices of animal sacrifices as a duty and virtue taught by the Vedas. This was the turning point of mankind's history. History chose to honor the gods of war instead.

In our teaching of the Vitalogical Sciences, we have based our discipline on Buddha's Dharma. Not only did Buddha abstain from flesh but he taught against eating seeds and raw meat, showing he saw a relationship of substance with reproductive enzymes to arousing sensuous desires of giving birth. He lived 6 years with only a hemp seed per day, becoming so weak he could hardly stand up, and yet the birth giving instinct remained with his flesh. Thus, he concluded, "There are two Extremes, O brethren Bhikkhus, which a holy man should avoid,- which are the habitual practice of self-indulgence, which is vulgar and profitless, and the habitual practice of self-mortification, which is painful and equally profitless." In this, in the Khaggavisana Sutta, he begins his discourse: "Having abandoned the practice of Violence toward all things, not doing violence to any one of them, let one wish not for children. Why wish for a friend? Let one walk alone like a rhinoceros...Men associate with and serve others for the sake of an object. Friends who have no object in view are difficult to obtain. Men are not pure. Let one walk alone like a rhinoceros."

There seems to be justification in holding that the Buddha spoke of St. John the Baptist, really meaning "Enlightened or Buddha in Aramaic, when he spoke of another Buddha, to be called Maitreya, who would appear 500 years after his Paranirvana. In The Surangama Sutra, foretold the conduct of John, the Light of the World, "Pure and earnest monks, if they are true and sincere, will never wear clothing made of silk, wear boots made of leather because it involves the taking of life. Neither will they indulge in eating milk or cheese because they are depriving the young animals of that which rightly belongs to them. To wear anything, or partake of anything for self comfort, deceiving one's self as to the suffering it causes others or other sentient life, is to set up an affinity with that lower life which will draw them toward it." One must take in account that "The Surangama Sutra" was written in Sanskrit by an unknown writer during the First Century A.D.", according to Dwight Goddard, who edited "THE BUDDHIST BIBLE" in 1938. This precept greatly influenced me in much of my 30 years abstinence from dairy products, as it did in my Incarnation as John. John the Baptist lived in the desert, where grass to tread upon is practically absent, where he could make his clothing of bark of trees as well as rope hemp fiber (rope hemp and camel hair are described by the same word in Aramaic) which also provided material for sandals, and his food grew wild on trees, such as carob pod and wild honey which may have also been a plant nectar of some kind. Thus, John surpassed Siddhartha among Buddhas, in avoiding injury to plant life, herbs, and milk, since the Siddhartha Buddha ate rice milk, destroying life in grains and milk by cooking and eating them. Pure and earnest Bhikkhus were limited to only one meal a day, but to avoid injury to herbs or animals, only fruit of plants is freely given by the plant.

Being a Life Member of the MAHA BODHI SOCIETY OF INDIA, 4-A, Bankim Chatterjee, Calcutta, 73, India, I can recommend their publication "The Maha Bodhi" as the best source of worldwide Buddhist news as well as excellent articles on principles of Buddhist Doctrine from both Northern and Southern Schools. They rightfully hold that Buddhism is the oldest historical and rational doctrine of spiritual endeavor. Likewise, a sapling of the Bodhi Tree (Ficus religiosa) under which the Buddha attained full enlightenment was brought to Sri Lanka (Ceylon) by Sanghamitta and it is surviving as the oldest historical tree in the world. In past years they have sponsored, not only planting this holy fig tree, but also they plant fruit and other trees as the positive emphasis of the Sanctity of all Life on earth.

To summarize on more of the Buddhist Doctrines of the

Dharma that has challenged all other conceptions worldwide, let us examine first THE FOUR ARYAN TRUTHS. The Four Noble Truths consist of self-evident facts: (1) The Universality of Suffering; (2) The cause of suffering rooted in desire; (3) By ending Desire, suffering comes to an end; (4) The way to end Desire and hence to end suffering, is to follow the Eightfold Noble Path. The word "suffering" means ill of any kind.

THE EIGHTFOLD ARYAN PATH consists of (1) RIGHT UNDERSTANDING or Right Insight or view, which is comprehending the Four Aryan Truths; (2) RIGHT RESOLUTION or Aspiration is seen in Renunciation, Benevolence and Kindness in all things; (3) RIGHT SPEECH is the abstention from lying, slander, abuse, and idle talk; (4) RIGHT ACTION, doing or behavior is abstaining from taking life, from taking what is not given, from carnal indulgence; (5) RIGHT LIVELIHOOD is a way of life that does not harm or exploit others, and exemplifies Virtue; (6) RIGHT EFFORT is effort made to maintain good states of mind, and effort to put away bad and evil states of mind; (7) RIGHT MINDFULNESS or meditation is remaining ardent, self-possessed and mindful of overcoming both hankering in desires and dejection in depressed feeling toward life, which comes with intuitive understanding of the Dharma; (8) RIGHT RAPTURE is fixing the mind on its pure essence, beyond word-thinking, transcending ready-made concepts, in which one abides in THE FOUR JHANAS or Holy States of Rapture (Dhyana or Contemplation): (i) Metta, Universal Loving Kindness to all, the unconditioned heart; (ii) KARUNA, Universal Compassion, heart-felt sympathy for all beings; (iii) MUDITA, Universal Joy felt in Thankfulness, Cheer toward all; (iv) UPPEKKA, Serenity in Peace and Equanimity of mind, transcending all dualisms thru Illumined Insight.

Explaining the originations of suffering and the phenomenal world the Exalted One stated: "On CONSCIOUSNESS depend Contact; On Contact depends Sensation; On Sensation depends Desire; On DESIRE depends Attachment; On Attachment depends Existence; On Existence depends Birth; On BIRTH depend Old Age, Death, Sorrow, Lamentation, Misery, Grief and Despair. Thus, arises this entire aggregation of Misery (that we experience in Life)."

The best starting point for attainment in Right Rapture or Dhyana leading to Absorption of Samadhi, is Right Breathing or ANAPANASATI. Anapanasati or Buddhist Breathing differs from Pranayama or Yogic Breathing of the Hindu tradition, in that Pranayama tries to Control this Vital Process, by imposing our will on Breath and consequently Mind, asserting our Ego Power. Rather than letting Cosmic Divine Will manifest, thru intellectual word-concepts, we promote a teacher or tradition which promises unlimited expansion of Self, still bound in selfish will seeking to fulfill our desires. In turn, in Anapanasati, one seeks to become conscious of this Vital Process, seeking to be One with its rhythm and Cosmic Participation. Breath is INSPIRATION!

If one takes to the Contemplative Life naturally, seeking to be one with the Cosmic Truth, Will or Divine Design, then Teachers and so-called "Masters" can only give a second hand experience of often traditional truths distorted by mythical concepts and word meanings. The Buddha claimed only to be a human being, sought no supernatural gods, followed no priests or teachers, and realized Intuitive or Spiritual Truths that anyone can realize for himself.

As we stated, in Anapanasati one becomes conscious of his Life Function as one breathes in and out. "We receive and accept the Vital Forces (Prana) of the Universe with our whole being with every inhalation, and we surrender ourselves wholly with every exhalation. This makes us realize that Life consists in a continuous process of taking and giving, of receiving and relinquishing, of integration and renunciation, of a continuous exchange and profound interrelationship of all individual and universal forces. Whatever we receive, we have to give back, whatever we try to hold on to or to keep ourselves, will kill us. "Whosoever shall seek to save his life shall lose it." In this concept we quoted Lama Anagarika Govinda's work "CREATIVE MEDITATION AND MULTI-DIMENTIONAL CONSCIOUSNESS", and add his Meditation poem "The Rhythm of Life", witnessing the in and out flow of breath:
"Threefold is the Rhythm of Life: Taking, Giving, Self-Forgetting. Inhaling I take the world within me, Exhaling I give myself to the world. Emptied I live within myself, live, without self, in Voidness Supreme. Inhaling I take the world

within me. Exhaling I give myself to the world. Emptied I experience Abundance. Formless I fulfill Form."

In this attitude of Spontaneity, consciously watching the inflow of Cosmic Forces and Substance into the body, the individual soon will contemplate that every part of the body BREATHES. The digestive tract breathes in gases in the form of Living Water composed of Hydrogen and Oxygen, etc. and expels wastes returned to fruit trees and the air. The Skin also breathes. The Heart sends out Love, Thanks, and receives Happiness. But most noticeable the Minds Respiration consists in the Incoming INSPIRATION of Cosmic contact we have with the Universe, and the EXPIRATON of our egotistical self and cravings. With this purification of being, the way of self-annihilation or spiritual death is avoided. Just to renounce the world, because we cannot accept or receive it, or we want only to receive without giving back what we owe to the world makes one not only physically, but spiritually mortal.

Lama Anagarika Govinda explains further that, the Secret to Immortality is in letting go of all our past, not holding on, not identifying ourselves with that which is perishable and impermanent as is the mortal body. The way to Samadhi is a continuous process of Spiritual Renunciation, a continuous giving up. Perfection is not an absolute value or static condition,-we cannot dwell in Samadhi indefinitely and Infinity can be experienced only by moving toward Infinity. As soon as we get accustomed to a condition of Samadhi, its Joy ceases. Light incessantly moves thru the universe but it becomes visible only when it meets resistance. In the same way consciousness becomes aware of itself only when there is resistance. If the resistance is impenetrable or insurmountable, it is felt as suffering; if it can be mastered it is felt as Joy. Joy means overcoming resistance. Life becomes worthless and unbearable if there is no resistance, or if the obstacles we meet cannot be mastered. The part of the Master, an adept of the Short Path is in helping the seeker to clear or rid himself of beliefs, ideas and habits in thinking, and guard against acquiring new ones; it is a discipline in avoiding the imaging of things and meditation is the rejection of thoughts with their seeds. For instance the conquest of death depends on overcoming that concept. If death were something real, we could experience it. But as long as we are to experience anything, we are not dead. Likewise, we "believe" in the existence of the sun; we see it, we feel it and we live in its light and warmth. Space is visible time, what is passed we see, and not what is coming. By its nature, the body is actually materialized Karma, the past made visible to the consciousness. The feeling of time is the feeling of Incompleteness. Space and time are the outcome of movement.

CHAPTER VII: THE HYPERBOREAN ORIGIN OF THE ROOTS TO WORDS, THE AGES OF MAN OR YUGAS, AND MAN AS THE COMPENDIUM OF LIFE.

Now that our Beloved Disciple, Student or Reader has the Heavenly Grace to diligently pursue study this far into the historical evidence of the cultural heritage that our forefathers derived from the Atlantians, and their antecedents, we are going to actually get involved investigating the cultures, Wisdom and the Word of the Heaven-Born Hyperboreans. When I say Word, this covers a lot of meaning, in that first of all the Word is Logos, the Wisdom from Above, and the Gnosis (Spiritual Knowledge) presented in a communicable symbology or language as well as phonetic sounds of speech. Connecting syllable sounds to syllable signs in writing enabled the Hyperborean Masters to deductively present syllogistic truths of the Living Logos. To become proficient in Hyperborean Speech was thus necessarily the discipline of orderly logic and reason which casts asunder all mystery.

In the Preamble of the "PRISTINE ORDER OF PARADISIAN PERFECTION", we refer to it as "THE AUGUST ORDER OF LIVING IMMORTALS "In Conscientious Tattwas",- to complete the original version) with ancient pre-historic foundations or Fountainhead at Kara Nor, that is Kara Usu Nor when the full name of this Sacred Lake on the North side of the Altai or Heavenly Mountains is given. Altho we prefer to speak of the original Hyperborean High Heavenly Hierarchy indication what others are used to referring as the GREAT WHITE BROTHERHOOD, LODGE AND ORDER, it has been referred to by ancient historians such as Herodotus, as the Hyperboreans simply. As we have shown earlier, the Hyperboreans had their cultural home of the Masters of Wisdom in the Heavenly or Altai Mountains which in early ages a 100 million years ago was a warm fructiferous Paradise in surrounding foothills, which of course was ages before the Gobi and Tarim Basin were fertile lands, and their flooding and eventual desolation into desert existing there now.

Due to the Genetic memory I inherit from having lived so many thousands of years ago, adding up to millions of years in experience with the Hyperboreans, otherwise known as the ancient TAEVASTAN or meaning Heavenly Finns of South Finland including Estonia, without confusing mutations of later racial divergence, the common source of all other racial, linguistic, and cultural traits became so apparent to me. The Key to this Divine Teaching, or the Pure Living Word (Logos) is found in its Living Tongue, which gives the basic meaning to the root languages of mankind, in spite of the confusion that the "moderns" worship in a Tower of Babel. The latter came by mischief created by tribes of man who purposely gave new meaning to words in opposite or misconstrued significance, sarcastic, sacrosanct or in blasphemy of the Sacred Truth, or bent to favor selfish purposes.

A complete dictionary would be a separate work in itself, and yet I would like to give an introduction, illustrating various basic words which with suffix syllables develop derived meanings in logical simplicity. Let us begin by taking the name of the mentioned Fountainhead of the Paradisians, or "The August Order of Living Immortals" at Kara Usu Nor. The Buddhist, Essene, Yogi, Hermetic, and other Orders speak of this common origin, reluctant to reveal their Source.

KARA USU NOR: In Sanskrit "Kara" means action, act, force, power, potency showing Being, Motion or Inherent Motive Energy. From this "Karma", the Law of Cause and Effect is derived. "Karya" is duty, and "Karuna" means Compassion or Loving Kindness. Then the meaning of such action (Kara) is also used for the means, which obviously is hand (organ of action in man), elephant trunk (organ of animal power) the sun's rays (solar energy), and thus author, cause and producer are included also among the definitions of "Kara". Phrenology characterizes human nose development with power and action, just as is seen in the powerful elephant trunk. Now, all this gets its origin in the Altai-Hyperborean language, just as all races came from that origin.

KARM, in Hyperborean means strict, sever, rigorous, disciplinary or harsh in action, like the Sanskrit Law of Karm(a). The Hyperboreans (or Hyprb.) "Karastama" means to harden as one tempers steel, to fortify, to toughen as well as refresh energies. "Karakter" in Hyprb. is the origin universally of the word

Character, referring to the legendary "grit" or steadfast discipline in virtue and will. Today we speak of the Science of Human Engineering as "Karakter" Building, but this goes back to the Altai or Heavenly Mt. Tribal Tongue that enabled mankind's cultural survival thru the Change of Axis of the earth in which their Paradisian Home turned to Glacial wasteland. Then, also "Karistama" means to discipline, punish or be harsh perfectionist in discipline that may be described as "Karske" or abstemious, while "Karsklane" describes the abstainer, teetotaler and perfectionist that the members of "The August Order of Living Immortals in Conscientious Principles" are known for. In turn, "Kargam" is to jump, prance, or romp as one does in triumph or fright, "Karjuma" is to shout or cry out, while "Kartus" is fear or anxiety, and "Kartlik" is timid or shy.

Having acquired the "Kar" prefix vocabulary, let us turn to the USU prefix which means belief. "Usk" (pronounced oosk) means faith, religion, belief and doctrine. Thus, "Usuline" means religious, "Usuteadus" is theology, "Usutunnistu" is one's creed, doctrine or profession of faith, and "Usu-opetus" means Divinity studies. "Usklik" is faithful or devoted, and "Uskuma" is the verb to believe. From there we have "Usaldama" meaning to trust or confide, "Usaldatav" or trustworthy or reliable, and "Usin" for diligent or industrious. Noteworthy herewithin to Gnostic origin research is the fact that "Uss (ooss is the word for serpent, snake or worm, and is symbolic for wisemen "Nagas" of India, Wisdom (Gnosis) the doctrine and truth of Gnostics, Ophites, Nassenes, Sethians, and reaches back to the Hermetic Caduceus of Greeks, Egyptians and Atlantians, all of whom received from Hyperborean culture. History, founded by Herodotus, connects mankind's cultural heritage as originating with these Hyperborean Paradisians.

The early Essenes, Estonians, Finns, Hungarians, Tibetans, etc. like their pre-historic founders of human culture, the "Karakter" Builders of Hyperborea had a devotional affinity to locate in proximity of "Sacred Lakes" and rivers. But why were such waters "sacred"? This again relates to a prehistoric devotion to cleansing, purification and baptisms, such as practiced by the Gnostic-Essenes on the Jordan, etc. including the First Christians of Saint John, as well as sacred tanks and Ganges water of India, the baptismal temple rites of the Egyptians, etc., by which body and soul were cleansed of sin, morally purified and restored to the blameless innocence of a young child. The Finns migrating to Europe chose "Suomi", the Land of a thousand Sacred Lakes as their home, and continued their Paradisian tradition even thru icy winters with their "Sauna", sweating out the body's impurities by improvised means dedicated to Helios as an institution comparable to temples, much like the summer's sun bath they took, unclothed and unashamed by tradition. After beating themselves with birch leaves, as they sweat out toxins in heat created by heated rocks, water cast upon them and enclosure of the Sauna temple, they found delightful and stimulating relief by bathing in a Sacred Lake (even breaking ice to enter) or rolling in the snow. One sees how this hardens the body, like heat and cold harden steel, giving the extremes that develop "grit" in "Karaktar". Consequently, this gave motive to Scandinavian-Nordic-Finnish tales of Father Frost Giant, or "Jattilainen" (Finnish: "Jaa" is ice, pronounced like "Ya") and the Abominable Snowman or "Yeti" of the Tibetans.

NOR is the word for lake from Tibet to Siberia, which is a derivative from the Hyperborean prefix "Noor" meaning young, "noorus" youth, "noorendema" rejuvenate, etc., because the lakes contained sacred waters of Eternal Youth, useful in cleansing, purifying and strengthening the body and consequently mind and soul. Water is like sunlight for giving life, health and youth. "Noor Yahee" is an Arabic derivative meaning "Light of Elohim", or Light of Divine Wisdom that is transmitted by World Teachers. In turn, "Norsu" was the Hypbr. term for the powerful snorting hairy elephant or mammoth of Siberia. Thus, Norisema means to snore, "norima" is to nag, carp or tease in mischief, while "norgus" means listless, depressed or dejected. "Norm" as now universally accepted means the rule or standard. "Norralane", Norwegian, comes from Norra or Norway.

From the above, KARA USU NOR as the Sacred Lake Sanctuary of the Hyperborean Guides of Mankind, it becomes evident this related to the Mount MERU of Hindu Legend. On the North side of the Heavenly (Altai) Mountains, the Sacred Lake was surrounded by the "City of Brahma" or rather the Abode of Heavenly Beings. Altho legend always speaks of it being north of the

Himalayas, on the first Polar Continent and thus directly indicates Hyperborea, yet this was spoken of being the middle or equator of the earth, which it was when the earth's axis was not yet placed so far north, and later became one of the 7 islands of the Atlantian Era. Kara Usu Nor is the Shrine of the Sacred Lake of the Almighty Faith simply, yet the attributes in Rejuvenation and Transcendent Spirituality of THE AUGUST ORDER OF LIVING IMMORTALS IN CONCIENTIOUS PRINCIPLES speaks of God as Almighty, and consequently the Source of All Life Everlastingly, that all may live, move and have Being.

In Hyperborean, "ELU" means Life; "Elu-kas" is animal; "Elustama" to revive, enliven or resuscitate; "Elujouline, vital; "Elukutse, one's professional calling or livelihood; "Elulugu, biography; "Elukoht", home or residence. In turn, "Elav" means alive or animated, but "Elama" is the verb to live, dwell or inhabit; "Elavus, a term for vivacity; "Elamus", experience; "Elatus, subsistance; "Elatama", to support, and "Elatanud", aged or elderly.

From the Hyperborean word "Elu" for Life, we get the Pre-Sumerian God, the Father, and from the Hyprb. "Hea" or Good, another Sumerian God. In the "Chaldean Account of the Genesis" in the Geo. Smith translation we read: "The Lord of the earth his name called out, the Father Elu",- and "The God Hea heard and his liver was angry because his (Angelic) man had corrupted his purity", so that Hea desires that "Wisdom and Knowledge hostilely oppose man". Later, this became the theme about the Tree of Knowledge, whose eating thereof, cast man out of Eden in the Biblic Genesis. God, the Father is "Elu" and the sons of God, co-Creators are the "Elohim". "Hea" or " Chrestus" (pronounced Christus in Greek) both mean "the Good" in Gnostic allegories.

In Book II of SECRET DOCTRINE, Stanzas from Book of Dzyan we read: "The men of the Lord of Wisdom are Immortal, and not the Lunar Ones... The Root of Life was the Ocean of Immortality, and the Ocean was Radiant Light, which was Fire, Heat and Motion. The Radiant Child of the two (Bright Space and Dark Space) shines forth as the Son who is the Divine Dragon of

Wisdom". Darkness is the Father-Mother and their Son is the Light. The Seven Creative Sublime Lords of Creation are the Esoteric Buddhist "Dhyan-Chohans", corresponding to the Biblic "Elohim", and Christian Hierarchy of Archangels. Seven is the great number of Divine Mysteries, Seven Eternities or Seven Breaths of the Dragon of Wisdom are the "Kalpas", great ages of time, during which the Seven Root Races appear on earth. "As the Heavenly Beings, the Spirit-born highest Dhyan-Chohan can but bow in ignorance before the awful Mystery of Absolute Being", which would require the culmination of conscious existence or Non-Being to realize, we might add.

Thus, the ELOHIM has evolved into meaning both the Almighty Creator, and the Gods of Creation in a plural form, showing the Pre-Sumerian and Hyperborean root was lost and given modified meaning. "EL" as we illustrated is the Hyprb. root syllable for the Power or God of Life, the Aramaic "Elohim" being only the plural of Sumerian Elu. Likewise, by phonetic modification "EL" became "Al" in Arabic, expressed most commonly in Ali. From these observations we clearly derive evidence that the Maha Chohan, the Father of Elohim, or the Sons of God, who guide each new root and age, and the "Java Aleim" is the Chief Lord of Hierophants of Sacerdotal Colleges and is the equivalent of Maha Chohan. "Noor Yahee" (or Ocean of Radiant Light), as mentioned above, is the Divine Wisdom transmitted by the Java Aleim. In the Scandinavian Eddas, NORNAS are the three Divine Brothers, "Urd" the past; "Werdandi", the present; and "Skuld", the future, all of whom relate to the Law of Karma. From these illustrations one may see that each tongue bends and modifies the meaning of these few words from their original significance, seeking to match new opinions.

Next, take the data about the original Hyperborean Continent that covered the Polar Cap, extending from Siberia to North Scandinavia, Greenland, etc. which in the Atlantian Age became divided into many islands. Due to ancient astrological allegories, this gave in to confusion with the 7 stars of the Great Bear, the 7 Great Planetary Spirits, said to be reborn on earth on 7 continents of the 7 root races, 7 Cradles of man, and so on and on. But it has had versions claiming 12 Silvery Island Dwellings of the Lords of the Zodiac in the blue Heavenly Sea, which were represented by 12 Islands in the Atlantian Age, the 12 Tribes of Israel, 12 Apostolic Patriarchs, etc.

The legend of Shambhala is most universal among Eastern Mysteries and yet much of the data has been bent to satisfy the interests of those who tell it. Often Shambhala is bound with Tibetan Tradition, and consequently it has often been linked with High Lamas, and these High Lamas, having acquired records and teachings from the true Shambhala Sanctuaries, hold that they are the very source. As we illustrated Tibetan Archives speak of Naacal Records which are dated from 70,000 to 100,000 years ago, but these records originate at Kara Kol, now renamed Przeval'sk, north of Tien Shan Mts. and southwest of the Altai Mts. in Siberia. Shambhala is the very center or nucleus from which all the Spiritual Wisdom of earth is directed. From their Great White Lodge, the Masters of Wisdom have guided mankind for hundreds of thousands of years. "The whole globe has convulsed periodically; and has been so convulsed since the appearance of the First Race, four times. Yet, tho the whole face of the earth was transformed thereby each time, the conformation of the Arctic and Antarctic poles has but little altered. The polar lands unite and break off from each other into islands and peninsulas, yet remain ever the same. Therefore, NORTHERN ASIA is called the "ETERNAL OR PERPETUAL LAND," and the Antarctic the "ever living" and "concealed"; while the Mediterranean Atlantic, Pacific and other regions disappear and reappear in turn, into and above the great waters." (H.P. Blavatsky). On this premise, as the land of the oldest cultural and spiritual records and habitation thru earth's convolutions, and the fact that the Cradle of the original pure white race, blond and blue-eyed fruitarians, was there, we can rest assured that other Great White Lodges or locations of Shambhala are only authorized or unauthorized associates. The Eternal Land was tropical, the Gobi and Altai Mts. regions were fertile fruit forests, and the Hyperborean race of Giants, just as these northern white peoples are the tallest compared to southern darker populations, even now.

"Behold," say the Elohim of Genesis, "man has become one of us." Thus, after the Sons of God, or Gods of Creation, came other

Elohim, "the Aleim", who were hierophants initiated into the good and evil of this material world. The Elohim being Gods or Powers thus became the prototype for the College of Aleim, Patriarch-Priests, whose chief of head was the "Aleim Java", and "Java" is God Male-Female, in that Jah means males and Hovah, Eva or Eve is Female, considering that Jehova has an alternate spelling as Yahve. The Zohar scripture held that in the beginning the Elohim were called Echod, or One, and this is the equivalent of "The Deity is one in many." This "One in Many" manifesting on earth were the Dhyani-Buddhas, Prajnapati, Amshaspends and Manus in Aryan scriptures, and this collective Logoi or Elohim are the Builders or Architects of Divine Ideation of things to be in the 7 Ages of man. These Elohim, Logoi or Dhyani-Buddhas were the Great Architects, Co-Creators that gave the Divine Design to Prehistoric Shambhala, the City of Brahm at the site of Mt. Meru, the Paradisian Hyperborea between four and five million years ago. This was the time of Krita or Satya Yuga, meaning the Golden Age. To understand concepts concerning Yugas, I am going to give an outline of the years adding up to each of Four Yugas.

I...	SATYA YUGA	contains 1,728,000 mortal years of humans.
II..	TRETA YUGA	contains 1,296,000 mortal years
III.	DWAPARA YUGA	contains 864,000 mortal years
IV.	KALI YUGA	contains 432,000 mortal years
	The TOTAL for 4 Yugas is	4,320,000 mortal years in one MAHA YUGA

In a Year of the Gods there is 360 years of mortal humans. Then, in turn 1,000 Maha Yugas constitutes a KALPA, or one day of Brahma, and equals 4,320,000,000 human years, or double for night and day of Brahma, so that 360 days and nights equals 3,110,400,000,000 mortal years in ONE YEAR OF BRAHMA; all this multiplied by 100 equal one MAHA KALPA, or Brahma's Age totaling 311,040,000,000,000 man years.

Mdme. H.P. Blavatsky assures us that these exoteric figures are accepted thru-out India, and dovetail with Esoteric works. Now, let us review scientific references which estimate the formation of earth to be in the Azoic Era 3000 to 5000 million years ago. Then came the Pre-Cambrian Era 1000 to 1500 million years ago, and after the life forms of rudimentary development of this time, come the invertebrates and marine forms of Paleozoic Era lasting 300 million years. The Age of Reptiles or Mesozoic Era lasted 130 million years, before the age of mammals and seed plants, called the Cenozoic Era covering 75 million years.

Fruit trees can be traced back to having existed 100 million years ago in the Cretaceous period in the latter part of the Mesozoic Era when flowering plants, deciduous trees and Sequoias became dominant. Fruit trees in the area of Malaya-Sumatra-Java are the result of long selection and cultivation. The breadfruit, bananas and pineapple, so widely found over the eastern world, do not seed themselves naturally but have to be planted in order to grow. Thus, Dr. Oakes Ames, research professor at Harvard estimated that man must have practiced plant breeding for about 500,000 years. The hand axe used to cut trees, roots, etc. and to plant much like we use a mattock, was estimated to have existed 500,000 years ago, to match the age of man, but now scientists are already up-dating mankind's age into millions of years from oldest fossil remains. This shows tree and plant culture tools have been made for millions of years, while the spears, arrows, etc. of hunters may go back much less. "Folk tales are haunted with remembrance of an ancient day when men lived in a Paradise of fruit trees. With the Hebrews it was the Garden of Eden, with the Celts, the lost island of Avalon; with the Greeks, that wonderful garden on an island in the Western seas, the Hesperides; with the Persians, the Haoma-Tree Paradise; with the Chinese, the garden of the Peach Tree Goddess..." (Henry Bailey Stevens) "All the available evidence," says Elliot Smith, "seems to point clearly to the conclusion that until the invention of the methods of agriculture and irrigation on the large scale practiced in Egypt and Babylon, the world really enjoyed some such Golden Age as Hesiod described. Man was not driven into warfare by the instinct of pugnacity but by the greed for wealth and power which development of civilization was itself responsible for creating. Stevens' "Recovery of Culture", again we reiterate, concluding, "Apparently it was early in the 2nd millennium B.C. that the main bulwarks of the ancient culture were broken. Then it was that the Aryans invaded Punjab from the north, the Hyksos gained control in Egypt, the shepherds spread over Palestine, and the barbarians sacked Crete." Much evidence has been presented against this.

"That primitive man obtained all he required as a free gift of Nature is a silly fable... Our age was not preceded by a Golden Age; and primitive man was absolutely crushed by the burden of existence, by the difficulties of the struggle against nature," argued Lenin, as presented in texts thru-out the Socialist world of Marxism. Moreover, our concepts of the spiritual origin in the perfection of an Integral Paradise, lacking nothing in potentiality of Eternal Life and Happiness, and the evident constant disintegration of matter into simpler elements as well as the degeneration of all species which were devolved from the original compendium of all life contained in perfected mankind, are not only challenged by Socialistic Marxism, but also modern capitalistic science teaching Darwin's theory of evolution. Engels, the con-worker with Marx, is supposed to have forwarded Darwin's theory of Evolution. Engels was of the opinion that our ancestors, fossil men-in-the-making, could not have become "complete" men without a meat diet, the acquisition and distribution of which facilitated the development of the social instinct. "A meat diet contains an almost ready state of the most essential substances required by the organism for its metabolism." Engels also regarded the use of fire as being of the greatest importance to the development of modern man.

When we observe a magician on the stage pull rabbits out of the hat of anyone in the audience, and catch doves out of thin air with a net, we may not be aware of how he does it, but we know rabbits and doves were not hidden in our hats or in the air, yet it is the Nature of the magician to be responsible for such things. But we do not say the magician's nature caused a rabbit to appear in someone's hat. Why does the nature of earth and all created things cause them to be created? Like the magician, Darwin's Evolution and Marxist Socialism evolve the whole Cosmos out of nothing and a state of chaos. By developing complicated formulas and theories about appearances in material things, they seek to prove that Science is the author of everything governed by their laws.

The hypothesis of Evolution collapses on the origin of life. No case of spontaneous generation has ever actually been observed. We are continually being told by the news media that Science, the Great God of Scientists, has created Life in a test tube; in every case follow-up articles show that no living organism was produced from inorganic chemicals, but it furthered possibilities of doing so in theory only. An amoeba consists of a 100 quadrillion atoms, principally carbon, hydrogen, oxygen, nitrogen and traces of others. The Evolutionist says this number of atoms, in correct ratio, ACCIDENTALLY met together, split from the existing compounds, and reassembled themselves into a living amoeba. With such Science Fiction preached by "Scientific Research," why not admit that the magician, by hidden magic, really was able to precipitate a rabbit in a hat or a dove from thin air? Why is that different from medieval authors claiming that the fruit of goose and sheep trees turned into geese and sheep, snakes were formed from female hairs in a moist environment, and worms, insects, fishes and even mice were generated spontaneously from mud, slime or filth?

The complex human cell of the human body alone contains genetic data that in regular type would require one thousand normal volumes to record. Norbert Wiener, founder of cybernetics, is used to rescue the Evolutionists: "It is not until the built in information has passed a certain point that the capacity of the machine for absorbing further information begins to catch up with what is intrinsic in its structure. But in a certain degree of complexity, the acquired information not only can equal that which has been originally placed in the machine, but can vastly exceed it...the really imposing and active phenomena of LIFE and LEARNING only begin after the organism has reached a certain critical degree of complexity." thus, in "Evolution of the Biosphere", M. M. Kamshilov concludes, "Self-reproduction of living things is therefore a function of their specifically organized complexity." All this is fine, but the machine, computer or man-made brain, had a maker, inventor and vast information collecting media for its appearance. A very complex mind, brain and living Spirit, Omniscient, Omnipotent and Omnipresent is the author of all these material manifestations. Writing on the same thesis, V. I. Vernadsky admits "Life creates in its environment the conditions favoring its existence." Life is the Creator, sustainer and transformer, and Nature manifests Life's Creation. Volumes of books with a complexity of theory and data are used to confuse the mind, capitalizing on common people's lack of such complex knowledge, to prove to man that man is an ignorant animal needing Science and Socialism to run man's life, just as religions have done before them.

It is claimed that when man's apish ancestors abandoned the fruit diet of anthropoids and the tropical climate, and adopted a stimulating diet of slaughtered flesh and used fire to cook food and warm caves or shelter with fire, it developed complex wrinkles in a brain that grew expanding man's skull. Thus, man's brain expansion enabled the skull to balance more proportionally on the vertebral column, relieving the great strain on muscles and bending the neck required of animals to see ahead and above, making the upright walking of man possible. For the proof of this the vast search worldwide was started to find the missing link that supposedly shows man evolved from apes and lower animals. Darwin was a vegetarian who became "fed-up" with Biblic fables on the nature of creation, and a God that approved slaughter of beasts, and thus supported the idea that frugivorous anthropoids were our "cousins" which should show man also is governed by frugivorous laws. However, Darwin and Scientific perversions of his doctrine, are in the dark as to how man and apes relate. They blind their vision to the fact that visible living humans are degenerating and de-volving before their very eyes. One of my students just sent me an article about a Chinese boy, Yu Chenhuan, who was born in Liaoning province, and totally baffled doctors by so much hair all over his body; his photo shows a hairy faced boy of 18 months. Stefan Bobrovsky, whose face and body were covered with long golden hair was known as the Lion boy. Mexican dancer, Julia Pastrana, died at 26 giving birth to a stillborn hairy son; her sister was also hairy. In India, a family was recorded in which father, son, grandson and female members of the family as well, were all covered with thick hair. As we said before, Krishna married an ape-woman.

Our textbooks today lament that more children are hospitalized from birth defects than all infectious diseases combined. Notre Dame professor, Dr. Harvey Bender, stated that, judging from the way mankind is experimenting ("horsing around") with Life,

"EVERYBODY WILL BE BORN WITH AT LEAST ONE OR TWO SERIOUS BIRTH DEFECTS IN A FEW MORE GENERATIONS." Yet, while mankind sees man destroying itself creating freak or beast-like forms thru genetically harmful mutations giving wild growths in tumors, cancer, disease and birth defects, it supports present living habits, claiming man evolved from such defective animal forms, justifying the means as scientific law! By diet, emotions, heredity, accidents during conception, toxic and damaging environment, etc. we cause the defects distinguishing the offspring from normal humans, and with an increased persistence these defects become race or even new species. Humans are continually being born with lower anthropoid traits, and it's likely that increased or "showers" of cosmic radiation, - just as man's use of atomic radiation, chemicals, X-rays, etc., - could have prepared widespread fossil evidence of what are called missing links between man and animals. However, when science reconstructs hair and flesh to supposedly match such fossil forms, most often with wild bestial looks, these are purely vivid imaginings seeking to prove man is a perfect beast. The sub-human beings could have been bald headed, hairless or with thick hair all over the body, as well as having many imponderable fleshy forms and skin colors.

   Our students realize, just as many scientific researchers are beginning to admit, that cooking and killing, not only destroy food value of living substance, but also make food harder to assimilate, due to a lack of integrating enzymes and vitamins that catalyze metabolism. But, Engels, in his ignorance of the healing effects of natural living food, claimed that the cooked flesh diet liberated man, since supposedly it is more satisfying, wholesome, easily assimilated, less bulkier than the vegetable foods, of less burden to jaws, teeth and digestive organs, and easier to obtain. His labor theory of anthropogenesis claimed that the distinguishing feature of man from animals came from this ability to fashion tools and use them actively and constructively in collective labor was specially tailored for Socialistic Political Goals, but facts betray these arguments by findings of other Soviet researchers.

   We have reviewed "THE ORIGIN OF MAN" by M. Nesturkh,

who in detail has analyzed dozens of theories of Anthropogenesis, to show the most acceptable scientific conclusions presumably, and expose the idealistic and religious conceptions as to the miraculous creation of the first men. As illustrated thru-out our writings, we have no special religion or concept of God to defend, we can live and philosophize even with the absence of any God who lays down laws in books, requires worship and rituals, and such anthropomorphism. Yet, in the theories of these materialistic scientists we find nothing so less dependent on the miraculous, and the preaching of complex unverifiable theories seems identically patterned to doctrines of theologians. As to our teachings on living foods and not depending on animal slaughter or raising of seeds for food, this we have verified in health obtained and sustained, healing illness of long standing, after being abadoned of hope by top medical scientists, altho we once were addicted to cooking and slaughtered foods. We do not preach an idealism which we have not realized in our Paradisian Life Style. Many of the suppositions entertained by the scientists, such as wild animals attacking man or destroying him, if he had not invented weapons, we find confined to carnivorous ways of thinking, since without offensive thoughts against carnivorous animals, poisonous snakes and ferocious beasts never attack a virtuous man.

Sinanthropi (Peking man) is estimated to have lived 500,000 years ago, which is before the time of Glacial Eras when ice sheets covered much of northern hemisphere, and when man was supposedly forced to use animal flesh and fire to prepare food. Yet Sinanthropi is believed to have formed bone and stone tools, and the charred bones and remains of 70 different species of mammals including antelope and deer, along with ashes 7 meters thick in his cave, are supposed to show he collectively not only made quartz weapons by chipping stone and sharpening bone, but also he feasted upon the roasted meat of beasts. Because of this meat-eating and tool making, the brain of Peking man had the average capacity of 1,050 c.c. as compared to his supposed predecessor, the chimpanzee, whose brain capacity is 350 to 400 c.c. The weight of man's brain is said to have expanded or increased to 10 times the weight of the gorilla brain, 4 times that of the chimpanzee, 6 times that of the orangutan and twice that of the gibbon, since man took to building fires to cook the meat he slaughtered with weapons to give greater concentrations of protein and phosphorous. The longevity of primates is increased compared to the majority of other mammals (excluding whales and elephants, etc.), but compared to big anthropoids who are believed to reach 60 or more years, modern man tops them all now with near 70 year average and reaching over 100 and beyond 150 in cases, all claimed for benefits of cooking, killing and working. The features typical to man which distinguish man are a very big, highly developed brain, hand with strongly developed and fully opposable thumb, ability to move on two legs, and foot resting fully on the ground.

In answer to these biast findings, we might bring out that two-thirds of mankind eat practically no meat, and have not done so as far as history can record, and yet they have no less brain capacity. The most recent findings claim that roasting meat (as supposed of Peking man) produces cancer-causing substances, as does cooking in general, the smoke and air pollution from homes, industries and cities greatly periling the health and existence of man. Engels admits, "Just as man learned to consume everything edible, he learned to live in every climate. He spread over the whole habitable world, being the only animal which had the power to do so on his own account." With a deeper understanding this also shows that abandoning man's frugivorous existence, the degenerating mind developed intellectually to pacify his protesting intuitive conscience justifying the eating of unnatural cooked food, planting grain fields where forest was cut down for firewood and building, which in turn produced deserts by erosion and exposure to climatic hostility. This pattern developed in the Gobi and spread westward all the way around earth, until now man has only outer space as the only place to travel to and spread his erroneous Life-Destruction Style.

Eating meat requires basically well over 15 times as much land as when fruits and vegetables are man's food, and even with the omnivorous diet including meat 5 times more land is required than for fructivorous existence. If the use of fire and flesh solved man's problems due to scarcity of food, it was because men destroyed forests. Man's problems would have been solved then, as they were from time immemorial, by building Paradises, providing fruit, shoots and leaves for food, forest to eliminate climatic hostility, erosion, and liberate man from the need of firewood, coal, petroleum, etc. to heat and cook. If man's brain developed such capacity due to killing, cooking and making weapons, there is much more capacity to be developed with the mass of today's ecological problems.

Paradises are not built by anthropoids swinging thru trees, hunting fruit, in spite of hard work, nor by idealists dreaming about them: They come by hard work with tools, and being able to enjoy them thru an un-cooked diet and not interfering with the animal life required for Paradisian Ecology.

The whole argument about brain size proves nothing, since the Indian elephant has a brain that weighs 4 times as much as man's brain. Human brains average from 1,200 to 1,600 c.c., but individuals with 800 c.c. brain capacity or less, are found to be in no way different from their fellows. Brain volume of many anthropoids,- the inferior to normal humans, shows greater intelligence in that they refuse to procreate in captivity, and stay close to a diet of uncooked fruit, shoots and leaves of trees. M. Nesturkh concludes: "If we take into consideration, the fact that the human brain varies within the limits of several hundred c.c., and that the level of intellectual development does not depend on the absolute weight of the brain, then no importance can be attached to this difference. A more logical solution is seen in the fact that the brains of some domestic animals atrophy, or lose size compared with their wild relatives. The brain of the goat has been reduced by about one-third. Since ancient records of historians, rather than speculating theorists, claim Paradisian man was much wiser than the immature reckless moderns, it could well be that the Paradisian Life close to nature develops the mind much better than the "domestic" civilized man captive to a life of tedious work for his food-package and box-dwelling. Without use the brain and body atrophies which can easily explain why lower anthropoids lost brain size and yet developed strong bodies after injurious life habits brought birth deformity to human families which transformed into increased tendency to their new life pattern. No longer would other normal humans breed with them so each generation increasingly favored built-in deformities and built up physical attributes of beasts undeveloped in humans.

"The History of life on earth is subdivided into six eras and a number of systems or periods of immense total duration equal to 3500 million years," writes M. M. Kamshilov in "EVOLUTION OF BIOSPHERE" and referring to a table by A. S. Vologdin. We must not question this since Darwin stated that the origin of the first living beings barred all new attempts of spontaneous generation of life! Science proclaims the doctrine so the world believes it. We did not say THEORY, because that contains "Theo" or God's prefix, revealing the miraculous. The geologic eras are divided into a geologic column of different earth strata which are labeled with ages ranging into thousands of millions of years. The age of each strata is determined by fossils of ancient life forms found within them. However, the sequence of the strata of rock layers is purely theoretical, or unproven, and no where are they found in complete in the order theorized. Often rocks of the most ancient eras are found on top of rock of recent eras, in reverse order, and complete confusion to which era preceded and which followed, so one might develop entire sequences in orders of varied patterns.

"The age of geological eras, periods, and epochs is now determined quite precisely by the ratio, in corresponding strata, of the amount of stable and radioactive isotopes in them", writes I. KARUZINA in a textbook on "Biology"... "Thus, the rate at which light uranium U235 decays to half of its original value is 713 million years, of heavy uranium U238 is 4,500 million years, of thorium (Th) is 13.9 million years, of potassium K40 is 1,200 million years, of radioactive carbon C14 is 5,600, etc." By "quite precisely" determining these ages we have witnessed some textbooks claiming one million years for the first man, Homo Pithecanthropus, while others place him at only 500,000, and fossils from the Mt. Carmel area in Palestine are claimed to be the first real man, who competes with Heidelberg man, Sinanthropus (according to Weidenreich) and several new fossil findings by the Leakeys in Africa dated in multiple millions of years. Carbon C14 concentrations in the air and sea have not remained constant over the years, scientists admit, but the theory made on many assumptions is better than none at all! They do not know or allow for the vast water canopy suspended high over earth before deserts appeared that would have shielded all fossils from cosmic radiation and other factors that could throw estimates off many thousands and millions of years. Potassium K40 and its decayed product, Argon, measurements are confused if traces of Argon remain in molten lava, giving figures that are far too high.

So, even the god "Science" may have got tired and had to rest the seventh day or a million years here and there. We hope geologic eras do contain some resemblance of fact, since the Biblic age of our earth is obviously only the tradition of a fanatic Jewish tribe, which recorded history even disputes. Yet, all these theories do not take into consideration that all inorganic matter was once living matter. Much new research is going on into how living organisms transmute one element into another, formerly cast aside as Alchemy or Shamanism, but now receiving the attention of advanced scientists. Plants made all soil and what we call elements are the corpses of living processes, is the teaching of Rudolf Hauschka, Baron von Herzeele and Louis Kervran. The matter that material scientists claim to be the origin of living organisms, is really the product of living organisms rather than the origin.

We shall not repeat explanations about how petrified living matter is the origin of rock and earth, which can be found in our course "The Lovewisdom Message on Paradise Building". But I would like to show that matter is in the process of DISINTEGRATION, while it is the Spirit of Life, Omnipresent, Omnipotent and Omniscient, that is the Creator, Sustainer and Transformer. Life was never an accident, but is the constantly functioning Law of Nature and Cosmos. The Creator is the Life and Nature of Creation. Material elements are disintegrating to particles of energy, Living Matter is disintegrating into Earthy Matter, Animal Life is disintegrating into simpler and simpler forms, and Humans are disintegrating into races and species akin to beasts.

The human fetus, after successful impregnation of the ova by the sperm, the simplest human reproductive cells build the embryo which becomes the fetus, is the witness of a microcosmic universe living within us, because the formation of a human being requires the guided direction of human intelligence thru all the lowest to the highest forms of life for the evolution of a potential Creator of the Universe. Contrary to Darwin's wishful belief, matter does not evolve itself thru all the simplest life forms to the most complex. This is why in animal forms the development of the fetus has not completely reached human intelligence in its original Spark of Life or Monad, and consequently is arrested in growth in its corresponding bestial stage of development, for its lack of complete humanity. The simplest forms of germ life are constantly being built by the ignorant human body causing a wild growth of embryonic cells whenever damage is perpetrated by toxins, etc., to consume the offense. This wild growth of trophoblastic tissue is the known cause of tumors and cancer, while my personal findings point to the even more evident fact that all disease organisms are nothing but the arrested growth of embryonic reproductive cells of the simplest kind created by human sub-conscious intelligence for the body's defense. As long as disease germs, tumors, etc. consume toxins, they cannot overwhelm the body poisoning it, but overtaxing this output of reproductive growths that consume the invading toxins, the body cells no longer are spared this poisoning and the body dies. Man's inharmony with Life and Nature thus gives birth to new organisms, the monstrous disease growths, which spread contagiously thruout man's environment. In the original races they converted the inorganic mineral substances into living plants to provide earthy food for man so man could have his physical vehicle, tangible to senses needed to develop human consciousness, rather than exist only Omnipresent, Omnipotent and Omniscient Void, Beingless Supreme Being. Such neutrality of attributes so overwhelms ordinary human mind, because the human consciousness requires the tangibilities to form mental concepts.

The Biblic Genesis described Eve as the mother of all that lived, moved and had being. The arrested growth of the human fetus was responsible for the birth of anthropoid apes when man violated the laws of man's frugivorous being. Human form evades disintegration by fulfilling a fructiferous diet of juicy fruits,- leaves and shoots of plants included. But as soon as humans began to use nuts and other seeds, the species took on the characteristics of Zinjanthropus, the Nut-Cracker Man, and other more ape-like fossils. From nut-eating apes with less of the living water in juicy fruits, and greater bulk in dry food substance in nuts and seeds led to development of the spider monkeys and squirrel monkeys, and then tree shrews, squirrels, etc. Other fossil men were probably egg-hunters and insect or locust eaters, and thus gave rise to the gorilla and other species that modified their form and structure to suit their lower intelligence of function. In time, fish, small animals and the fare of carnivorous beasts, the instinctive habit of karmically less-evolved living monads, appeared.

The dietetic violations, starting with the use of nuts and seeds, not only developed menstrual or oestrous cycles in humans and fossil man, but chimpanzees have a cycle of 36 days, the gorilla of 45 days, the orangutan of 32 days, and monkeys of 31 to 42 days. In turn, the fact that elephants have such large bodies, and a brain 4 times larger than man, shows that before the arrested growth in size of humans, the giant humans of legends probably did exist with equal dimensions. The atrophied male breasts show female origin, pointing to an androgynous origin of the early races of man. That some humans have capabilities of higher spiritual development, and specifically, many develop the pineal or third eye, shows that originally the species lived truly as Gods.

FUTHORE - THE ANCIENT HYPERBOREAN ALPHABET

The runestaves (or letters) of the FUTHORE (the first 6 letters of this most ancient Nordic alphabet, their a-b-c's) originally meant something secret, mystical or magic. Run is Old English for whisper, secret counsel, mystery, etc. The Finnish Runos (Poem-songs) are used by the ancient bard, Vainamoinen to musically vibrate molecular essence, or create and transform things into being. What is meant is a God-Spell (origin of Go'spel), scriptural incantation engendering faith and being. GOTHIC (Scandinavian), SAXON, GERMANIC, ENGLISH AND CELTIC (Edge carving) alphabet letter symbols.).

GOTHIC RUNES

ANGLO-SAXON RUNES

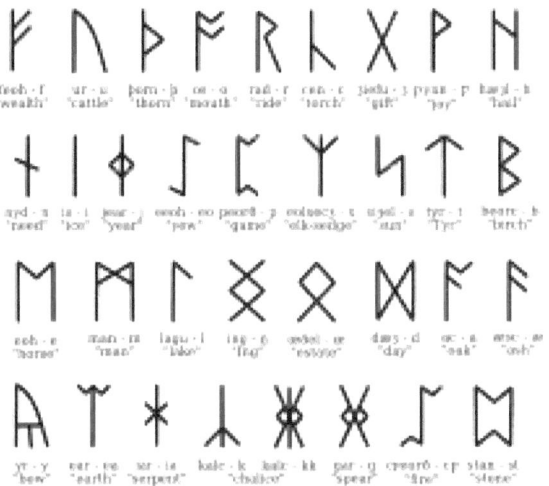

MEDIEVAL RUNES (12th to 15th centuries))

ᛁᛒ ᛚ ᛏᛑᚦ ᛂᚠᚠ ✱ ᛁᛈᛐᛦ ᛕ ᛁ ᛒ ᛔᚱ ᛁ ᛉᚿᛘ ᛜ ᛚ ᛂ ᛡ
abcdþðefghiklmnopqrstuvyzæø

CELTIC RUNES

Symbol	Name	Meaning	Letter
ᚁ	Beithe	birch	b
ᚂ	Luis	blaze/herb	l
ᚃ	Fern	alder	f
ᚄ	Sail	willow	s
ᚅ	Nion	fork/loft	n
ᚆ	Uath	fear(?)	h
ᚇ	Dair	oak	d
ᚈ	Tinne	rod/metal	t
ᚉ	Coll	hazel	c
ᚊ	Ceirt	bush	q
ᚋ	Muin	neck	m
ᚌ	Gort	field	g
ᚍ	nGéatal	wounding(?)	ng
ᚎ	Straif	sulphur	z
ᚏ	Ruis	red(ness)	r
ᚐ	Ailm	?	a
ᚑ	Onn	ash-tree	o
ᚒ	Úr	earth	u
ᚓ	Eadhadh	?	e
ᚔ	Iodhadh	?	i
ᚕ	Éabhadh	?	ea
ᚖ	Ór	gold	oi
ᚗ	Uilleann	elbow	ui
ᚘ	Ifín	pine	ia
ᚙ	Eamhancholl	double c	ae
ᚚ	Peith	soft birch	p
᚛	Elte	feather	
᚜	Spás	space	
	Elte thuathail	reversed feather	

Notice that Hyperborean tongue origins did not refine sounds, as modern tongue do, so one letter served the purpose of 2, 3 as well as one sound letter of modern alphabets. They communicated by True-Thought or Extra-Sensory-Perception, so Truth was felt instantly, the simplest syllable sounds being only needed for clearing perplexity of those whose perception lacked purity in the medium. Runologists have now traced Rune from earliest Gothic origins in Scandinavia (from whom Germanic, Saxon, etc. developed) to Finno-Ugric peoples, with samples in Hungary and the farthest ancient Siberian-Hyperborean fountainhead. Thousands of Runic inscriptions reveal that the origins of Greek and Phoenician alphabets go back to the Cradle of Mankind in Northern Asia. The first Gothic-Scandinavian runestaves (5th century B. C.) read from right to left, used 16 to 24 letters, and certain letters (runestaves) represent 2 or 3 English letters now in use. Runestaves from Lapland to the Danube, and from Ireland (if not America) to Siberia, show that the spoken word and written language was of One Source, Hyperborea. Celtic letters were carved on the edge of stones, boards or staffs, thus covering more than one plane, arcane.

Facial skeleton of baboon too heavy for lacking neurocranium to support head up, and balance seen in human skull.

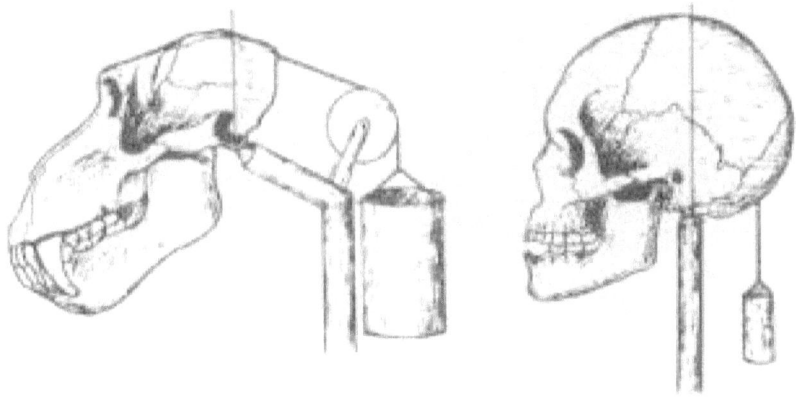

Mystical Anthropology

Kraken is a legendary sea monster of giant proportions that is said to dwell off the coasts of Norway and Greenland. Doreal's statue had two ancient Gobi humans with the Kraken (small one) being held by one, while the other meditates with hand over his pineal eye. (Chapt. VIII)

Above one sees the ugd peaks of the Heavenly or Altai Mountains, practically being impassible, which gave Hyperboreans the reputation as having mt. goat's feet.

Mystical Anthropology

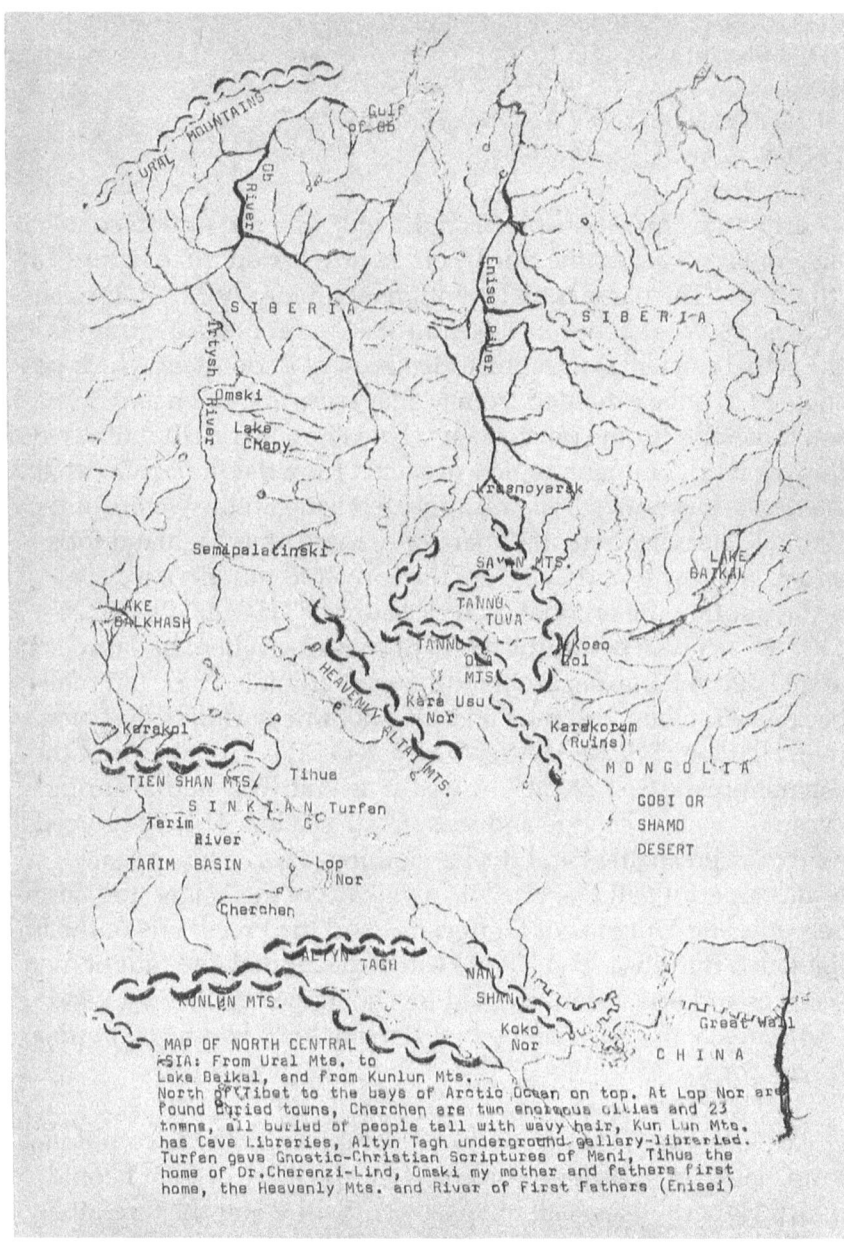

MAP OF NORTH CENTRAL ASIA: From Ural Mts. to Lake Baikal, and from Kunlun Mts. North of Tibet to the bays of the Arctic Ocean on top. At Lop Nor are found buried towns, at Cherchen are two enormous cities and 23 towns, all buried with people tall with wavy hair, Kun Lun Mts. has Cave Libraries, Altyn Tagh underground gallery-libraried. Turfan gave Gnostic-Christian Scriptures of Mani, Tihua the home Dr. Cherenzi-Lind, Omski my mother and fathers first home, the Heavenly Mts. and River of First Fathers (Enisei)

CHAPTER VIII: THE FIRST PEOPLE, THE LEMURIANS, THE GOBI CIVILIZATION, ANTHROPOGENESIS OF MODERN SCIENCE AND MYSTICAL LEGENDS ABOUT SHAMBHALA

After the "Imperishable Sacred Land" and the second continent of Hyperborea, the third Continent we propose to name "LEMURIA". "The name is an invention, or an idea of Mr. P. L. Sclater, who asserted between 1850 and 1860, on zoological grounds the actual existence, in prehistoric times, of a continent which he showed to have extended from Madagascar to Ceylon and Sumatra. It included some portions of what is now Africa; but otherwise this gigantic continent, which stretched from the Indian Ocean to Australia, has now wholly disappeared beneath the waters of the PACIFIC, leaving here and there only some of its highland tops which are now islands." To get the correct foundation, we are again quoting the original esoteric source, "THE SECRET DOCTRINE" as given by Mdme. H. P. Blavatsky. "When the Third separated and fell into sin by breeding man-animals, these (animals) became ferocious, and men and they became mutually destructive. Till then, THERE WAS NO SIN, NO LIFE TAKEN. After the separation, the SATYA YUGA was at an end. The eternal spring became constant change and seasons succeeded. The cold forced men to build shelters and devise clothing. Then men appealed to the superior Fathers. The Nirmanakaya of the Nagas, the wise Serpents and Dragons of Light came, and the Precursors of the Enlightened Buddhas. The Divine Kings descended and taught men sciences and arts, for man could live no longer in the First Land (Adi-Varsha, the Eden of the first Races) which had turned into a frozen corpse."

"The Hyperborean word "eest" combined with the combining form "lane" or "Eastlane" means "First-Lander" or First People, which Hyperboreans call themselves. "Esi-isa" means forefather, while the word "Eden-eme" means to thrive, flourish or prosper, since "eme" is mother, while "maa" is one's land as well as earth, and "edasi" reveals meaning as forward or onward movement of Elu (Life) on Maa (earth)."

The "Egg-born" Third Race of Lemurians fell, and created no

longer, it BEGAT its progeny. Being still mindless at the period of separation it begot, moreover anomalous offspring, until its physiological nature had adjusted its instincts in the right direction. Like the "Lords the Gods" of the Bible, the "Sons of Wisdom," the Dhyan Chohans, had warned them TO LEAVE ALONE THE FRUIT FORBIDDEN BY NATURE: but the warning proved of no value. Man realized the unfitness,- we must not say sin,- of what they had done, only when too late: after the angelic monads from higher spheres had incarnated in and endowed them with understanding. To that day they had remained simply physical, like the animals generated from them. The doctrine teaches that the only difference between animate and inanimate objects on earth, between an animal and human frame, is that in some the various "fires" are latent, and in others they are active. The vital fires are in all things and not a thing, or an atom, is free of this (atomic) fire. But no animal has the three higher principles awakened in him; they are simply potential, latent and thus non-existing."
All this is the Secret Doctrine as taught by Helen Blavatsky, and reluctant as I am to promote special doctrines, it is so delightful to find such corroboration to my teaching as to eating forbidden food, seeds and flesh, causing bestialization of man giving origin to animals.

"As the embryo of man has no more of the ape in it than of any other mammal, but contains in itself THE TOTALITY OF THE KINGDOM OF NATURE, and since it seems to be a 'persistent type' of life, far more than even the Foraminifera, it seems as illogical to make him evolve from the ape as it would be to trace his origin to the frog or a dog." She laments that not only do modern scientists reject statements such as, that of all mammals, man was the earliest, so man is the indirect ancestor of the ape, and that he had a giant body in his earliest appearances. "Nor can they admit that the first two races of men were TOO ETHEREAL AND PHANTOM LIKE IN THEIR CONSTITUTION, ORGANISM AND SHAPE, EVEN TO BE CALLED PHYSICAL MEN. For, if they do, it will be found that this is one of the reasons why THEIR RELICS CAN NEVER BE EXPECTED TO BE EXHUMED AMONG OTHER FOSSILS. Man was the storehouse, so to speak, of all the seeds of life for this Round, vegetable and animal alike. "As soon as man appeared, everything was complete, for everything is comprised in man. He unites in himself all forms."

"The mystery of the earthly man is after the mystery of the Heavenly man", means that the human form is the vehicle of the Divine Creator of Creation on earth, or its Nature, we repeat. This gives one of the 7 keys to interpreting the Story of the Ark. In the Mazdean "Vendidad" we read, "Thither (into the Vara) thou shalt bring the seeds of men and women of the greatest, best and finest kinds on earth; thither thou shalt bring the seeds of every kind of cattle, (etc. ...All those seeds shalt thou bring, two of every kind, to be kept inexhaustibly there, so long as those men shall stay in the Vara." Those "men" in the Vara are the Progenitors, the Heavenly men or DHYANI, the future Egos who are commissioned to inform mankind. For the "Vara" or the "Ark" or the Vehicle simply means man.

Even more curious are claims that in the third Race man became atrophied in former attributes: There were 4 armed human creatures in the early days of the male-females (hermaphrodites) who were with one head, yet 3 eyes. They could see before and behind. But a Kalpa after the separation of the sexes, men having fallen into matter, their spiritual vision became dim and the 3rd eye started to lose its power, became petrified and disappeared. The double-faced became one-faced, and the eye was drawn deep into the head, now buried under hair. It is also claimed man was a transparent creature, like the chameleon, when that eye was open. The 4 armed Hindu gods, the "eye of Siva" and androgynous qualities, are all related to the Lemurian Mysteries of an ancient origin when India and Sri Lanka (Ceylon) were parts of the Lemurian continent. The 3rd Eye is indissolubly connected with Karma. This is why Chelas seeking higher Spiritual Truth must inescapably observe Celibacy or Continence. When Androgynous man began to propagate thru animal acts of sex, the Siva Eye became atrophied, and Karma followed man in all his actions. It was the atrophy of the Third Eye that still prevents perception and knowledge of Reincarnation and its Karmic Cause.

Soviet researchers have also become interested in the almost forgotten legends of Lemuria, a vast continent that vanished into the Indian Ocean. This came into evidence when researchers were excavating in the Turkmen Soviet Socialist Republic. These excavations brought to light an advanced civilization very much older than "Urartu", a well studied ancient civilization of that region. Some 5,000 years ago, those "lost" people had cities, temples, forts and astonishing production processes. Their works are said to resemble the Ubaid-Sumerian people of Mesopotamia. Thus, these findings seem to show, Lemurian colonists may have travelled from the Indian Ocean continent, from Java and Madagascar right across India to Mesopotamia and up to the Siberian region east of the Caspian Sea. Being west of the Pamirs and Tien Shan Mts., once called West Turkistan, it was the home of Aryan Celts. Other evidence that the Soviets gathered included ancient India scripts mentioning Lemuria, data from Arab Geographers of olden time describing islands completely lost today which may have been the remaining parts before complete disappearance. Their language clues include the fact that East African languages closely resemble the Dravidian speech of ancient India and the Bay of Bengal region. To our research it accounts for how it came that the Paradisian Hyperboreans were introduced into eating cooked flesh and bread, to become the Warring and Callous Aryans who later invaded Europe and India. What links this widely separated region are the Mysteries of TANTRIC RELIGION. What is called the Oldest Religion has two characteristic deities, Siva and his Wife. We have already mentioned the "Eye of Siva" beginning to atrophy in the Lemurians, that evidence was found at Harappa and Mohenjo-daro in the Indus Valley, and of course, Siva, the Transformer, is part of the Hindu Trinity. However, Soviet researchers also found that Tantrism had been long taught in Buryatia, the Lake Baikal region. Herodotus gave historic credit to the Hyperboreans for their claims about the earth being a globe or round, but Soviet scholar Pubayev finds that Tantrism also taught that the Earth is a sphere rotating on its axis. These ancient people had 4 wheel vehicles, and made use of the boomerang weapon of Australia.

Now, the slant that these researchers are presenting is that all Civilization had a common origin in a Lemurian Garden of Eden, because they are slighting the fact that Siberia was once warm and tropical and they search for remains of an Eden that they have right on their doorstep in Siberia. But they have found pictures on rocks of prehistoric extinct animals, and among them weird creatures called "Wondjinas" in Australia. Their legends were thought purely imaginary, but now Wondjinas are identified as resembling those mysterious men from Space found in many places in the world. Not only in Australia, but in Tassili in the Sahara desert are such pictures found complete with Astronaut helmets.

In 1936, I contacted the LEMURIAN FELLOWSHIP, which moved from Wisconsin to Chula Vista, California and later Ramona, Calif. where I visited them in 1938. It had a great attraction to me then, telling of lost cities and economic idealism. California was claimed to one extreme of a vast continent that reached across the Pacific to Australia. They claimed to have knowledge and wisdom thru 78,000 years of experience by the Lemurian Masters. However, now it is 26,000 years since the continent of Mu, as they called Lemuria, sank beneath the blue Pacific, seemingly gone forever. So they preached a renaissance of Lemurian Culture, holding faith that the lost continent would rise again as it was prophesied by them. Possibly shocking to Bible students believing Adam to be born in 4,000 B.C., they held Melchizedek was Emperor of the Great Mukulian Civilization. Melchizedek was Christ's first appearance 50,000 years before he appeared as Emperor Poseidonis of Atlantis. The revelations are said to be from Akashic records, but one soon sees the Bible as their principle source of idealism, and brief research into ancient ruins of buildings of unknown origin across the south Pacific.

We now have come to a clearer perspective to understand the great importance I hold for the study of the Hyperborean Paradisians and the true Genesis of Satya Yuga of Spirit-Born Humanity, which will reveal on what principles there might be the Restoration of that Golden Age. While doing intensive Spiritual Study with the GREAT WHITE LODGE during my hermit venture living in a grass hut on Lake Quilotoa (1945-1949), among numerous courses by various Gurus whom I personally received instruction from, Dr. Doreal gave much data about Shambhala, the Ancient People of the Gobi, etc. In fact, the first credentials to practice Spiritual Healing and a Minister's License came from Dr. Doreal's College, incorporated in Delaware and Colorado, for training Chelas of the Great White Lodge. They established an esoteric retreat high in the Rocky Mts. surrounded by Mountains containing lead deposits to act as a shield against atomic radiation and disaster now reaching toward climax. My tuition was arranged thru Master K. H. and Anagarika N. K.

Records from Shambhala of the Great White Lodge, which D. D. gave illustrations from, told of a time when the GOBI was a great fertile tropical country, a time which goes back as far as FOUR AND FIVE MILLIONS OF YEARS AGO. Thus, it is that the Gobi is one of the great centers from which all the civilizations, the races of mankind, migrated and were scattered to the far ends of the earth. "Of course, I am familiar with the great migration theory of India as being the home of Aryan Culture. However, that was a later migration than the one I am going to speak of, so let us discuss the ancient record that has been preserved in the great libraries of Shambhala," he explained. The natives say the Gobi was the home of all races in the beginning. They trace legendary history hundreds of thousands to millions of years ago. Occasionally some very strange objects have been brought from their ancient cities, bronze images, utensils and gold amulets with engraved characters hard to decipher.

Of course, many of the ancient cities are only 1,500 to 2,000 old. Paintings, carvings and inscriptions show they were of a white race resembling the Aryans. These peoples were some of the earliest followers of the Buddha, evidently long before the Tibetan Lamaseries. However, of more interest are archaeological findings that date back to the age of the dinosaur. D. D. claimed the only dinosaur eggs that were ever found, were found by Chapman in the Gobi. He (D. D.) dated the great continent of Lemuria as having existed 600,000 years ago, contrary to Lemurian Fellowships dating at 78,000 to 26,000 years before it sank.

"Anthropology claims man is of comparatively modern origin," D. D. would say, "yet footprints of man have been found imbedded in rock which geologists say is over TWO HUNDRED AND FIFTY MILLION YEARS OLD." Then he explains that at that time, the Gobi was not the desert it is today, but instead the most fertile spot on earth. It was threaded with streams and four

great rivers that came from surrounding mountains, for which reason some say the Gobi was the original Garden of Eden. Now, there are only dry river beds, but in the winter when it rains, the water pouring into those old river beds brings
up all kind of objects made of gold, silver and bronze. For many hundreds of years the natives have used these streams for treasure hunting, which shows a great many people once inhabited this area. "The ancient records say that the race which grew up in the Gobi was a fair-skinned, blue-eyed and blonde-haired race; that was the FIRST OR ORIGINAL RACE", is a statement that matches the meaning of "Eastlane" of my Hyperborean tongue and origin.

Even up to 550,000 years ago the earth was still covered with a great water mist; the sun did not shine as it does now; and the world itself was shrouded in a mist so heavy that human beings moved thru a dense fog, but to compensate, they had powers and vision that man does not have today. The Third Eye was active at that time, and it looked out at the top of the head, where the soft spot is in the baby's head. A thin membrane covered it, and it also had the power to see on what is called the infra-red range of vibration, so they could see thru the misty fog as well as you can see now in full sunlight. This misty canopy covered to the height of over 300 miles... Then D. D. continued explaining that at that time, all mankind still had gills in the throat like fish do today. What are now called tonsils, were particular organs of the body which would draw in that water mist with the breath, separate the water mist from the air, and allow the water to come out thru openings behind the ears.

Like the Shaman native legends still claim, the Gobi people could transmute earth and common metals into gold and silver, were able to move physically in the 4th dimension, and knew how to release the power of the atom, D. D. claimed. They had a symbol which looks like what is called a devilfish today, and it represented the Cosmos. They called it "Kraken", because it had tentacles that were powers and forces that reach out and grasp and hold all things in their place. This symbol is found on rocks, in caves, etc. in the mountains of the Gobi, and in the Scandinavian countries of Europe. In the Gobi for tens of thousands of years there was no war, but only a great peace.

With the gradual increase of population in a limited area, they began migrating to distant regions. After the races and tribes had migrated from the Gobi to Atlantis and Lemuria, one tribe remained, or the last part of one tribe. That was the fair-skinned, blue-eyed, yellow-haired race that had existed there for thousands, or rather, millions of years. They lived there until a period of about 4,000 years ago. They left the Gobi Desert because there came an influx of barbarous races from Mongolian countries. That race carried with them some of the ancient legends. They migrated to parts of Asia and then, into Europe and they formed what we call the Scandinavian races. The Great Kraken still appears in Scandinavian legends, but it was sacred to what is called the "Green Race" by D. D., and from it was the Hammer of Thor, which was later used as a weapon being a piece of metal with metal tipped tongs attached which was thrown to catch prey. Dr. Doreal claimed that he had the oldest statue in the world, a priestly symbol from a temple excavation of one of the most ancient cities of the Gobi. In it there are two persons, one is holding his hand over his Pineal Eye, to show he has gone within meditation, and the other holds the Kraken far above his head. These people are dressed in grass skirts tied with a rope, their beard is curved out and their hair is drawn out in a peculiar hair style, and they rest upon the base of the statue which is a rock embedded with sea shells of an unknown fossil type, while the curved design seems to represent breakers of the ocean. This status is dated at 50,000 years ago, believed sometime after the sinking of Atlantis and Lemuria, and in D. D.'s opinion (like presented in an earlier view herewithin) the people of the Gobi were not destroyed with the sinking of either of those continents tho they knew of it. The figures in the statue are not of bronze, tho there is much copper in the metal used to implant a tremendous and powerful vibration. Placing the statue so one touches the top of the Kraken and also the top of the head of the other figure, a sensitive person can clairvoyantly perceive the odic force in vision of what surrounded the image when it was made thousands of years ago.

At this point, I wish to pause, and go into other idealistic hypotheses of anthropogenesis, related to this study. A prominent American paleontologist, Henry Osborn (1857-1935) assumed that the Tertiary progenitors of man in the form of Eoanthropi, or

Dawn men, had almost all the qualities possessed by modern man, and had no connection with the ancestors of apes. Osborn journeyed with Charles William Andrews to the Gobi Desert which served as an impetus for this elaboration of the ETERNAL CHARACTER of the human type with its original habitat in Mongolia and Tibet. During the expedition to Mongolia with its mountain plateau landscapes and abundance of strange, extinct reptiles,- dragons to dinosaurs,- Osborn conceived the idea that it was precisely here that the cradle of man was to be found; he pictured it as a region with winding streams and scant forests alternating with grassy plains. In 1929, Osborn developed his hypothesis into a widely based theory of the semi-arid plateau in a paper he read at a congress of the 200th Anniversary of the American Philosophical Society. He conducted a polemic against Darwin's contention that it must have been a hot climate with heavy tropical forests that man developed in, and thus accounts for the development of long legs for hiking, development of tool-making, thumb, etc. Osborn's theories include Aristogenes, meaning better heredity and adaptation to the environment in orthogenetic development.

A prominent German anthropologist, Franz Weidenreich, proposed the Gigantoid hypothesis of the origin of man in 1945. Its factual basis is the fossil remains of huge anthropoids found in South-East Asia, which Weidenreich regards as hominids (human) and the most ancient representatives of the human race. He placed importance on the general changes of the body being observed in the dental system. About 1,000 teeth of the Gigantopithecus have been found. (Tung Ti-cheng 1963) The crown molar of Gigantopithecus averages 4,420 cu.mm, that of a gorilla is 2,365 cu.mm., but man's 3rd lower molar only has the volume of 723 cu.mm., which means that the Gigantopithecus or Giant Man had teeth 6 times greater than modern man, and likewise, a body 6 times as large. Judging from the lower jaw with a full set of teeth (Pei Wen-chung 1957), the strength of Gigantopithecus was superior to all other species of animals, and had no need to take recourse to weapons to defend himself. Another support to Weidenreich's hypothesis is the lower jaw of ancient Javanese Meganthropus which likewise was of giant size, also indicating a huge body. Also in India there was the giant Dryopithecus, in South East Asia the names ranging from Australopithecus, Ramapithecus, Sivithecus,

Punjab Dryopithecus, to Megaladapis and Pro-consul major giant lemurs in East Africa, and all of which tends to substantiate our contention that they were from Lemuria. Weidenreich thus held that our ancestors were in the nature of overgrown Adam and Eve. Altho Osborn and Weidenreich are accused of supporting religious motives in demonstrating a hypothesis of degeneration, reduction in size etc. yet the agnostic theorists base their arguments on miraculously evolving everything from nothing, and science has only given mankind the most threatening causes for further degeneration and extinction of the human species. The teachers of evolution gave us only wordy theory.

The above theories and teachings that were in vogue in the 1920's 1930's and 1940's show their psychic influence on various teachers I then studied with, which was combined by the Western teachers with early Sunday School themes from the Bible. My good fortune was to have evaded childhood "Christian" indoctrination, so that when I did study the Bible I was emotionally ready, and I did it weighing everything with equal emphasis on doctrines from other religions, so I did not think of everything in dogmatically Christian terms, just as one is limited by one's childhood tongue to express everything according to its grammatical possibilities. Altho not as limited as the Lemurian Fellowship, Dr. Doreal's concept was a Shambhala of N. T. Apocalypse origin partially.

However, there was much to the good, and because he was dealing with a highly indoctrinated Christianity, he was forced to speak in terms of what they could understand. "The Brothers of the White Lodge, the Elder Brothers of Mankind, the Guardians of the human race, are those who have MASTERED LIFE and FREED THEMSELVES FROM BONDAGE TO THE WHEEL OF LIFE, and being free they can work and manifest and live on ALL PLANES OF DIVINE BEING/...The Great White Brothers of the Great White Lodge have as their work, as their destiny, living in material bodies to aid and help and instruct mankind in spiritual development. The Brothers of the Great White Lodge number 144, the same always. If one of you should attain to Adeptship and should be called into Hidden Retreats of Shambhala, you would be worked with and instructed by one who is fulfilling a particular destiny.

That one would pass his work over to the new Adept and go to another world where he is needed, so there are always 144 Adepts in the Great White Lodge."

"Shambhala is the very center or nucleus from which all Spiritual Wisdom of the world is directed. The Great White Lodge on this planet earth is the guiding intelligence, guiding the power of mankind. The Buddhist monasteries and Lhasa are only outer veils of the inner monasteries. (D. D. wrote this before the Lamaseries of Tibet were invaded). "Those who have passed thru Esoteric Buddhism are finally admitted into Shambhala just as there are many others who have passed thru many religions in the world. The Great Ones take no account of religion, as all men are seeking to find God...There is no religion higher than Truth, and in the last analysis there is no religion,- there is only God. The path never makes any difference." D. D. goes on to describe records about ancient happenings from the Gobian Age, the Atlantian Age, the Serpent People, and there may be found the record of everything that mankind ever did. Spark-like fireflies seem to appear in city lights that reflect the consciousness of every human soul on earth and someone is always watching these lights.

From this you find a modernized version of the Great Book of Life told of in the Apocalypse, where all the sins of Christians are recorded, which in turn are described as Akashic Records in Eastern Philosophy. Dozens of White Brotherhood Schools dramatically invent all kinds of devices said to record the spiritual conscience of mankind. D. D. uses the A.D.A.M. formula to arrive at the incarnations of Messiah which he describes as Adam, David, Aaron and Melchesideck. He said that Jesus was a Nazarene, because a Nazarene is an acolyte of the Essenes, Nazarenes being neophytes of the Essenes. The town of Nazareth did not exist until 350 A.D. when Empress Helen sent commissions to find the places mentioned in N. T. Bible allegories.

Now concerning Maitreya, the Lord of the Word, Gautama Buddha said to his disciple Ananda: ("The Blessed said to Ananda), "I am not the First Buddha who came into this world, A Holy One, a Supremely Enlightened One, endowed with Wisdom, auspicious, embracing the Universe, an incomparable leader of Men, a Ruler of devas and mortals who will be another Buddha will arise in the world. He will reveal to you the same Eternal Truths which I have taught you. He will establish His Law, glorious in its origin, glorious at the climax and glorious at the goal in the Spirit and in the letter. He will proclaim a righteous Life, wholly perfect and pure, such as I now proclaim. His disciples will number many thousands while mine number many hundreds." Ananda said, "How shall we know him?" "He will be known as Maitreya, Lord of the world." The year of my Initiation for Leadership work in the Spiritual or White Lodge (Pujili 1942) was followed by Doreal's teachings on Maitreya and Shambhala. Shambhala (correct spelling, and not Shamballa) represents the ideal state towards which all mankind is striving. "According to eastern legend, Shambhala is variously, a city, a valley, a temple, a monastery or a state of Consciousness; we might say that it is all and none of them." Explanation is given about Northern Shambhala and Southern Shambhala, and is described in prophetic terms, symbolic of his mission. Concerning the Dalai Lama and the Tashi Lama, Dr. Doreal wrote, "The Dalai Lama is actually the soldier and political ruler of Tibet, and because the coming of Shambhala threatens his reign, this is from the inside, you have heard of the TASHI LAMA having to leave Tibet, or at least disappear, but the Dalai Lama at the present time is the one spoken of as being the last Dalai Lama, the last of the political rulers; and Lhasa, the great city will fall, will be obscured and destroyed..." All of this came to pass, and also the prophecy "To those in darkness, the Banner of Maitreya shall flow as blood over the lands of the NEW WORLD." In the Hindu work "The Kalki Purana" we read, "At your request I shall take birth into the abode of Shambhala. I shall create Satya Yuga and restore its former condition, and after destroying the serpent Kali, I shall return to my own abode." Dr. Doreal lamented that the consciousness of India is now far below the West and that the Avatar had begun his work in America.

In the book, "TIBET", its history and beliefs are explained by the brother of the Dalai Lama, Thubten Figme Norbu, said that, "In our ancient scriptures we hear of A COUNTRY TO THE NORTH, THE COUNTRY OF SHAMBHALA, where there is to be a final battle between the forces of religion and atheism."

North of Tibet is Sinkiang (China), Mongolia, and the Altai Mts.,beside Siberia. Exactly straight up from Lhasa along the 91 and 92 degree Longitude North we locate KARA USU NOR surrounded by the snow capped Altai Mts. to the south and west, and the Sayan Mt. to the north and Lake Baikal on the east. Norbu also says, "Shambhala is described as a LAND RINGED WITH SNOW-CAPPED MOUNTAINS, and at its center is an enormous city in which the king has his palace. Some say that it was here that the GREAT LAMED TANTRA ORIGINATED, for king Suchandra, the first of Shambhala's priest-kings, took the teachings direct from Buddha. Up to the very end the capital city of Shambhala will be one place where the teachings of Buddha are preserved....Buddhism teaches us that our ignorance is suffering, and we know it, but that little spark of knowledge even brings beauty into our lives, helping us to see beauty everywhere, teaching us wisdom." Norbu now lives in the U.S.A.

The above gives important clues to the Northern Shambhala. Now, to introduce our Southern Shambhala, and Maha Maitreyana Mandala work, I must refer to the Sungma Red Lama. He has published a book "AGHARTA" in which he tells the mystifying tale of the Atlantians escaping from their palace on Mt. Olympus, before their continent sunk, by journeying in flying saucers to new homes in the gold palaces of Shambhala, the capital of the subterranean world of Agharta. Russian explorer F. Owssendowski claims tunnels encircle the earth, which were built by a race of supermen, possessing superior powers in the sciences, and came from a Hyperborean civilization which flourished in the polar region when that land was still tropical. The people of Agharta live in the hollow interior of the earth, where legends hold is a "Paradise of great beauty to the north, a land of perpetual youth, where tall, strong and handsome people live for thousands of years." I first heard of such a Hollow Earth Paradise when travelling with Prince Arlo in 1938, and later investigated many sites where tunnel entrances were to be found in the High Andes, entrances to the Inner World.

However, Sungma Red Lama, wrote an earlier book, "THE ETERNAL FOUNTAIN" in 1947 describing also the Southern Shambhala. "Signs of MAITREYA": "In the Era of Maitreya, the

World Teacher, -NOW,- flowers shall bloom in profusion and out of season and the Woman of the species called Mankind shall wear them, besides clothes of contrasting but harmonious colors. The music and dances of South America shall become predominant. THE COUNTRY NOW NAMED ECUADOR, S. A. WILL BECOME THE SPIRITUAL CENTER, THE TIBET OF THE WEST, after MAITREYA, the World Teacher has established himself over all the Earth under the Banner of Maitreya,- which is Truth and Justice for all Earthlings." He introduced this revelation by another statement quoted from another of my early teachers: "Maitreya as seen by the Lord Maha Chohan Kwang Hsih, H. H. O. M. Cherenzi-Lind, quoting from his Wesak Message of 1934: "The Lord Maitreya is not fully manifested and let it be understood, will never be a person, altho the sense-deluded mystics and the anthropomorphists of religion will give him a human form, with whiskers and all. The Lord Maitreya will only be a cosmic or universal aim and urge, endeavor and value, principle and realization,- not a man, a messiah or a Divinity." As we have shown elsewhere, Maha Chohan Cherenzi-Lind represented the flag of the Great White Lodge, or Banner of Maitreya as that of LOVE-WISDOM. Behind the mystifying claims for Maitreya (also meaning Christ and Kalki Avatar) who is to come riding a white horse to lead his forces to Victory, there is no person, because one (each person) must overcome the self (ignorant attachment to the impermanent). "Precious jewels on Maitreya's ring are each soul that is a vessel of infinite light, as taught in the doctrine of Bardo, and they encircle Maitreya (Love-Wisdom Consciousness) in the central Blessed Land (Shambhala) in the Sacred Land of Agharta (Inner Paradise, or Heavenly Kingdom). Anagarika N. K. wrote me once that the Tashi Hutulktu Prince Cherenzi-Lind "came to New York in 1933, where the Great Shambhala for the deceased Dalai Lama was celebrated." From this we see that Shambhala is much like the Christian Heaven or Paradise. Jesus of the N. T. says,- "This day thou shalt be with me in Paradise", and shortly they lay him in a tomb, as the sense-bound worldlings see it, but that cave of his tomb was like so many sacred caves extending into the world within our earth consciousness. Andes Indians insisted that the bat filled caves entering Tolas had a secret passage into the mountain where one sees a beautiful garden with delicious fruits, vases filled with gold, and other wonders.

Just as bat dung gives off gases that can intoxicate the mind, when one travels in the Arctic snows numbness gradually takes over, and people who freeze show a state of bliss on their face. We do not doubt that visits to a Polar Paradise entering within hollow earth are not real experiences, but the Giant Hyperboreans, many times as large as themselves they meet, are on a plane beyond the physical. We all know of people who experience visits from people who passed on, and feel such a spirit world is as real as the physical world. However, the Lovewisdom Message, or Maitreya, that I as a living being teach, has nothing to do with many things told about a physical "Maitreya" or a "Shamballa" of the deceased "spirits". Shambhala, the Kingdom of Heaven within may be realized outwardly in Building Paradises and living the Divine Design manifest on earth, as Love and Wisdom exemplified.

After the era of Mdme. Helen P. Blavatsky, who made THE SECRET DOCTRINE the Great Classic and Scripture of Theosophy and Esoteric studies in general, in the early 1920's, Alice Bailey founded The Arcane School and The Lucis Publishing Co. giving out a teaching I first learned of when traveling to see all the Californian cults in 1938. Many of her viewpoints have become guideposts characterizing my teaching, altho I studied almost nothing of her many works. She describes matters we are discussing now by saying, "The central home of the hierarchy is at SHAMBALLA, a centre in the Gobi Desert, called in the ancient books the "White Island". It exists in etheric matter and when the race of men on earth have developed etheric vision its location will be recognized and its reality admitted...The hierarchy of Brothers of the Light still exists and the work goes steadily on (since Lemurian days). They are all in physical existence, either in dense physical bodies, such as many of the Masters employ, or in etheric bodies, such as the more exalted helpers and the Lord of the World employ." Alice A. Bailey (or A. A. B.) then explains that Sanat Kumara, the Ancient of Days and LORD OF THE WORLD with an etheric body, is the Youth of Endless Summers, the Fountainhead of the Will, showing forth as LOVE of the Planetary Logos. The 3 main groups or divisions in the Hierarchy under the Lord of the World are the Manu, the World Teacher and the Maha Chohan. The Manu sets the racial type. The World Teacher is the Great Being whom

Christians call Christ, and is known in the Orient as Lord Maitreya and to Moslems he is described as Iman Madhi. He it is Who has presided over the destinies of life since 600 B.C. and He it is Who has come out among men before and Who again is looked for. He is the great Lord of Love and Compassion, just as his predecessor, the Buddha, was the Lord of Wisdom. In turn, the Maha Chohan is the Lord of Civilization, and his work is the fostering and strengthening of the relation between spirit and matter, life and form, the self and the not-self, which result in what we call civilization. Thus we have WILL, LOVE and INTELLIGENCE as three LORDS.

THE SOLAR HIERARCHY, The Solar Logos form the Solar Trinity
I. The Father in Heaven is known as the Divine Will
II. The Son is known as Love-Wisdom
III. The Holy Spirit is known as Active Intelligence
THE SEVEN RAYS: 3 Rays of Aspect: Will or Power, Love-Wisdom, Active Intelligence; 4 Rays of Attribute: Harmony or Beauty, Concrete Knowledge, Devotion or Idealism and Ceremonial Magic.

      THE PLANETARY HIERARCHY, The Lord of the World, Sanat Kumara. As the One Initiator he presides over Three Kumaras of Buddhas of Action.
I. The WILL aspects= The Manu; under him Master Jupiter + Master Morya
II. The LOVE-WISDOM aspect= The World Teacher or Christ and Maitreya, under whom a European Master, Master Kut Humi, Master Djwal Khul or Tibetan are included therein.
III. The INTELLIGENCE Aspect= The Maha Chohan or Lord of Civilization, under whom The Venetian Master, Master Serapis, Master Hilarion, Master Jesus and Master Rakoczi work with.
"The Lord of the World, the One Initiator, he whom the Bible called "The Ancient of Days and Hindu Scriptures the First Kumara, He Sanat Kumara it is, who from His throne at Shamballa in the Gobi Desert, presides over the Lodge of Masters, and holds in His hands the governments in all three departments." In regard to planetary centers we have:
I. SHAMBALLA (The Holy City) = Will or Power, Sanat Kumara, Melchizedek = Planetary Head Centre, spiritual Pineal Gland. Purpose=Plan=Life

II. THE HIERARCHY (New Jerusalem) = Love-Wisdom, World Savior or Christ = Planetary Heart Centre. Group Unity=Consciousness

III. HUMANITY (City standing foursquare) = Active Intelligence, Ruled by Lucifer, Son of Morning Star, The Prodigal Son = Planetary throat Centre. Self-Consciousness=Creativity.

Now, to prevent confusion from such brief outlines, may I add that Hercules, the Sun-God, had a First Ray Soul, 2nd Ray personality and a 6th Ray mind. The Christ had a 2nd Ray soul, 6th Ray personality (relationship with Jesus) plus a first Ray mind. A. A. B. added, "Very definitely may the assurance be given here, that prior to the coming of the Christ, adjustments will be made so that at the head of all great organizations will be found either a Master, or an Initiate..."

The above outline does give a precursory glimpse into the mystical or esoteric origins of the government of the Spiritual Kingdom of Mankind. Both Alice Bailey and I have received gifts from the Master Kut Humi, beside Initiations and I knew he was a person having a personal body from a visit we had in Quito (see Autobiography), but being entrusted with continuing his work and my Mission within the Solar Logoi, I learned more about him than the hazy picture these Theosophic folk could give.

It is ridiculous either to expect western observers to know the Arcane Secrets of the Hierarchy in the Gobi and of Shigatse, Tibet, or even get the real truth about them from the worldly political bosses of Tibetan Theocracy as given by the Dalai Lama and other well known sources. The Flight or the Secret Removal of the Hierarchy, from the Gobi and from Shigatse, Tibet is found mentioned thru-out esoteric circles, but like the location of Maitreya, and the Hierarchy, it can only be experienced on the etheric plane. The Prince of Ch'an, Oung Maung Cherenzi Lind, Ven. Tashi Hutulktu Kwang Hsih, and also the Tibetan Master Kut Humi Lal Singh, as well as the ex-Anagarika Lhasshakankrakrya are the worldly aspects of the Great Maha Chohan's one identity who transferred the World's Spiritual Center to the Andes. The seal of the Hierarchy which we use on our works, of course, is not absolute proof, since the Heavenly Hierarchy inherits the Eastern Wisdom, but has the Innate Mission of New Age Revelation direct from Solar

Hierarchy sources. Pr. Cherenzi-Lind spent many years at positions he had in Tihwa, Sinkiang, the western terminal of the Gobi Desert trade route, knew the ancient legends and modern findings in the Gobi, Altai, and the Cradle of man. Likewise, Shigatse, Tibet, called him in the days confusing Tibetan spellings, named his post Dashe See Gompa, Chygatzeh, Thibet and the world spiritual centre was labeled "Maha Bodha Mandala". He described how upon coming to the west, he had at first worn his beard, turban, and also items of his office in Tihwa and Shigatse which breathe out the smell of yak dung, used in the building, to heat shelters in high, cold and arid areas and which was repugnant to the western worship of self-purified ideal of the Hierarchy. Many of his descriptions of his work and his environment were unique, revealed aspects of the Arcane Hierarchy of a hoary Antiquity. In a lecture delivered by the Ven. K. H. in Los Angeles, Jan. 28, 1935 on the HIERARCHY he stated: "The Hierarchy cannot be composed of people who are alive, unless such members of the Hierarchy would have something to do, something to accomplish, a sort of mission in life, some sort of experience they would have need of, some sort of example to give to mankind, and only in that case would they come here to life as a human being, as a living incarnation of their own teaching." The Hierarchy must deal with life problems.

CHAPTER IX: ALL MATTER ONCE LIVED, HYPERBOREAN TRAITS, MAN'S EARTH ARRIVAL, DESCRIPTION OF ALTAI REGION, AND SACRED SANCTUARIES

With the sudden rapid development of materialistic sciences in the last few centuries, man's consciousness became crystallized into habitually thinking in terms of exact sciences, as if nothing which cannot be weighed, measured or counted did not exist. However quantitative and medianistic, world conceptions found no cause at this origin, and thus theories and hypotheses of what physically was incapable of proof were evolved, based on hypothesis and logic. The law of the conservation of matter has been regarded as one of the most fundamental laws of nature, which thus assumed atoms and particles composing them are eternal. The discovery of radium suddenly smashed this comfortable premise of scientists because it disobeyed the law of the conservation of matter, since radium disintegrated into electricity, warmth, light and various substances such as lead, helium, etc. Thus, all solid ground known to man's physical senses was again shattered, and human beings were after all but the chance product of swirling atoms and electrons, and no longer could truth be looked upon as a constant reliable fact now that science became the explorer of chance origins and speculated eventualities. Science-Fiction became of age, the most comprehensive calculations of Science being accepted as an ever changing theory of fact, and its variations considered futuristic novels of what may become fact for coming generations.

Goethe held that light was a far loftier element than the sphere in which its waves manifest, just as man's eternal being is the spiritual cause of the human anatomy. Novalis, in turn, wrote, "Man was reduced to a mere creature of Nature. The eternal creative music of the universe became the monotonous rattling of a giant mill driven by the stream of chance and floating on it, a mill sufficient unto itself, without a maker or a miller, a real perpetuum mobile, with only itself to grind!" Goethe's followers, such as Preuss, held that Matter and Spirit are a unity, and matter is Spirit on another level or plane. These findings brought to light experiments of Baron von Herzeele, who published 500 analyses showing an increase of potash, magnesium, phosphorus, calcium and sulphur content of seeds sprouted in distilled water only.

Reviewing these findings, checking their accuracy, Rudolf Hauschka in "THE NATURE OF SUBSTANCE" declared: "Thus, it appears not only that plants can transform substances, but that the creation of the basic elements of matter is commonplace in the organic kingdom." Herzeele went so far as to affirm that dead matter is never the primary origin. "What lives may die, but NOTHING IS CREATED DEAD." "The soil does not produce plants; PLANTS PRODUCE THE SOIL". The supposed immutability of chemical elements is now the out-of-date Science-Fiction of the past. "It will become evident later on that all the chemical elements known to science at the present time are more or less in the category of waste products,- CORPSES OF ORGANIC LIFE".

Diamonds are pure carbon, the HARDEST as well as most shining of earth's substances. Millions of chemical compounds found in nature all started by carbon combining with itself (as found in grape sugar) and with other elements to form carbohydrates, made possible by plant life giving form and substance to invisible gases of our atmosphere, carbon dioxide, oxygen and hydrogen.

Hauschka found that the whole earth was all one living organism. The mineral content of seedlings was observed to change according to changes of the earth's position in relation to the moon, sun and astrological signs of the zodiac with rhythmic rise and fall. "Light and darkness create the rainbow, earth and cosmos create the living kingdom of earth." Goethe's concept of two poles and a third new element, rhythm, which reconciles these two extremes, gave a clue as to why the ancients, by intuitive truth, gave astrology and alchemy importance as the basis of life on earth. "The earth does not manufacture soil by some physio-chemical process: it is the plants that create the soil by coming into material manifestation out of the universe. Not only does it create the fertile layer of humus over the underlying rocky earth in the course of a long creative life activity: it makes earth in a much broader sense," Hauschka explained. Coal and peat moss are fairly well known as deposits of plant life. Primeval plants were in a delicate and fleeting process of using rarified elements of the atmosphere. Some mountain areas are known to be made entirely of animal remains,- mussels, snails, etc.,- which form ammonite chalk.

Limestone is likewise the remains of the skeletal parts of animals, while primal stone is of vegetable origin. Silica is formed from the remains of bird feathers which are 70% silica, and the sense organs and animal skin are largely silica. In petrified wood, as found buried when digging wells, etc. is seen the visible evidence of how plant life gives fiber that will eventually age into stone in the earth. Clay (aluminum silicate) gives origin to rubies, sapphires, emeralds, etc., all of which are formed from plant and animal carcasses. Iron is found both in chlorophyll and the blood, showing how metals even come into being by a life process. Nerves are built of protein, phosphorus, etc. just as bones give origin to calcium, phosphorus, etc. all of which are derived from plants. Phosphorus has a greenish glow in the dark, and like electricity it is able to change oxygen into ozone giving the characteristic odor. Likewise, magnesium has affinity to light, and mountains having magnesium in the form of dolomite limestone give off a gentle rose red glow after sun set. The magnesium originated in the chlorophyll of plants which are made of sunlight, or whose fibers and substance are materialized light.

We have spoken of the highest evolved beings having taken formless bodies of vaporous substances, water being the key "alkahest" of alchemy for transmuting of substances in living things. From the above one may soon realize that such highly evolved beings, or the spiritual aspect was thus intelligently responsible for channeling light chemical rays of cosmic origin which eventually took on solid form of substance with the rotation of the vast heavenly sphere, of which earth is an evolving particle. Hauschka explains that in the Patonic year (25,920 sun years) the vernal point completes one round of cosmos. The vernal points advance from one constellation to another and has marked correspondingly the advance from one cultural period to another. We now start the Aquarian Age having finished the Piscean Epoch; the Aries epoch being the Greco-Roman Age, while Taurus or the Bull characterized the Egypto-Chaldean culture, which in turn was preceded by the Gemini culture of the Persians, prior to which was the Cancer Epoch of India, and all of which was preceded by the Leo-Sun Worship Culture of the Pamirs-Tarim-Gobi region. "Cosmic order weaves and pulsates thru the whole of Creation descending step by step to the mineral phase of earthy substance. The creative spiral permeates every phase of events; the spiritual, the psychological, the biological, the mineral." (Hauschka) "Silica minerals are products of Aries impulse, Sulphur and its relatives are of Gemini, Sodium and other alkalis of Virgo, Lime and mineral alkalis of Libra, aluminum and its brothers of Capricorn and Halogens are of Pisces. In turn the vernal point drops back a little so in 2,000 years it passes thru one constellation in the opposite direction. This accounts for the formation of Nitrogen by Taurus, Phosphorus by Cancer, Hydrogen by Leo, Sodium by Virgo, Carbon by Scorpio, Magnesium by Sagittarius, Aluminum by Capricorn, and Oxygen by Aquarius, all of which are tempered by different planetary spheres of creative impulse.

Earth, like Heaven, is thus shaped by a musical ordering power of the Cosmic Word, which John speaks of in his Gospel, the earth being God's Word materialized. In the esoteric interpretation it speaks of the Heavenly Elohim, the Spiritual Co-Creators of Paradise, and emphatically the First and Second Races. With a cosmic Origin of Life on Earth, and the emanating tones of vibration that symphonically patterned the form and elements that life used to clothe this living earth, we must also accept a Cosmic Purpose operating with Omniscient Intelligence with which to guide its course, rather than unthankfully crediting it to mere chance occurrences. We might imagine how Vainamoinen, "the bard immortal, sprung from the Divine Creatrix of the virgin Ilmatar, the Air-Born, as he melodiously composed, sang and played his Finnish harp to give creative vibrational composition to earth's first Paradise. "A mighty hero, while his feet the earth was stamping, to the clouds his head was lifted, to his knees his beard was flowing, to his spurs his locks descended." Once the tree of evil or darkness was leveled, after 3 mighty steps from ocean to shore and inland, "The clouds extended widely, the rainbow spanned the heavens, over the cloud-encompassed headland, and the island's misty summit, the wastes were clothed in verdure, and the woods grew up and flourished, leaves on trees and grass in the meadow, in the trees birds were singing, in the tree-tops called the cuckoo. Then the earth brought forth her berries, shone the fields with golden blossoms, herbs of every species flourished, plants and trees of all descriptions, but barley would not flourish..." (Kalevala) Thus the Finnish bard and farmer tells of it.

Now, to contemplate more of the nature of these Giant Paradisians, of which the "Kalevala" epic was a poor modernized picture, and regardless of our esteem for the present human form prejudged by familiarity and attachment to it, they surely must have been a giant species, able to handle the needed function of keeping and creating the First Paradise on the Hyperborean island. Rather than seeking to discriminate, or seek out the good and evil and other dualities, making two sides or dimensions visible as in the two-eyed bestial beings in animalish self-centered perspective of all things, these Gods-Goddesses were selflessly dedicated in their single purpose of doing God's will, or intuitive Divine Design in action, and thus could depend entirely on the Pineal Eye, evading the clashing aspects of matter which produce pain. The "National Enquirer" recently carried the following data on atavism or the reappearance of ancient traits in humans: BABIES BORN WITH ONE EYE IN MIDDLE OF FOREHEAD LINKED TO CYCLOPS LEGEND: That legendary one-eyed person called Cyclops may actually have existed, and in fact there may have been whole tribes of them, says an expert, Dr. Rufis Howard points out that even today some babies are born with one eye in the middle of the forehead,- and he thinks these deformities could be the result of still-surviving genes from ancient tribes. "Just about all cases of cyclopia are caused by genetic reasons. It is possible that a cyclop's gene could have been carried on in a latent hidden state. Dr. Richard Neu, geneticist at Buffalo Children's Hospital said one eyed people are a very rare occurrence, But it does happen, about one in every 40,000 births is a cyclops, according to Dr. Howard (clinical Prof. at Yale dept. of ophthalmology and visual science), A cyclopic child has an eye in the center of its forehead, with a fingerlike projection, kind of like a primitive nose (if humans had gills, as Doreal held, such a nose only protected the eye, -Ed.) above the eye, he said. Instead of having normal eyelids, it has a diamond-shaped opening and can't close its eye. The phenomena has been observed and documented, it seems, for thousands of years. "Scientists have found some cuneiform tablets along the Tigris River going back about 4,000 years which describe a cyclops being born that long ago", he revealed. One family in Europe had 5 cyclopic children."

We also have another clipping showing "BEARD! 17 and half feet long displayed in the Smithsonian was grown by a Norwegian immigrant over a span of 51 years". Thus Vainamoinen's beard betrayed him not to be a youngster with a beard to his knees clothing him like a skirt, 17 ft. being over double his body length, but considering Vainamoinen's stature reaching from above the earth's cloud canopy, from heavens to earth, the beard and tale must have been Titanic!

The mention of a finger-like proturbance like a nose above the eye sometimes called the Horn of Wisdom, or the Intuitional Bump sought out in phrenological analysis in the upper mid-forehead, probably protected the ever-open eye, as well as suggestive of the "All-Seeing Eye" of Divine attributes. The fact that the pineal eye arises in animals also, shows that the first animals were likewise endowed with the Intuitive Eye, meaning that altho some beings descended to animalistic existence, yet they recognized the holy ones or gods, and even today animals seem to sense who will harm them causing them to prepare for attack. The claim about the Serpent race once inhabiting the southern extreme of the globe, relates to the existence of the Serpent eye, traditional Kundalini Eye, 140 million years ago, and that man resembled Saurians described as having fossil skulls with a perforation indicating a much developed eye in the forehead. Reptiles with diamond-shaped heads are poisonous generally, and under a skin covering in the pineal region they have a third eye that senses danger from those who would harm them, just as with cases of the peaceful existence of poisonous snakes and humans I have heard of and experienced myself. Very often such snakes are found under the head of my bed, among the boxes of literature we send to people, especially the skin they shed is to be found wherever they molt or have been seeking to control rodents. One Sunday as I sat reading Scriptures on a stool with back to the door, a poisonous macanchi entered under the door and stool, crawling up my leg to the knees of my lap, and then went off to a nearby railing to the middle of the room. I knew better than move, bothering it when it was on and near me, but at a distance, I no longer was feared an offence to it, but seeing me rise, the snake began rattling as do rattlesnakes. Since man has become a killer, two eyes of discrimination must be used also to protect beings, so altho my aura was peaceful and clean, yet the serpent could not trust often betraying actions. The serpent and a child are portrayed in innocent co-existence.

The preceding first part of this chapter we devoted to establishing principles that make it easier to envision the existence of our heavenly origin. The elements of earth are constantly being formed and transformed by plant life, rather than plants originating from earth minerals. The weight of all the plants is 10,000 times greater than that of all animals, and together, plants, animals and people or all living things on Earth is approximately ten million tons. But if we were to calculate the total mass of the living organisms that have populated Earth thru-out the 4,500 to 5,000 million years since life first appeared on it, we would get a figure many times GREATER THAN THE ENTIRE MASS OF OUR PLANET. ("Biology" by T. Karuzina) This fact alone proves that the growing and living earth is the source of seemingly dead earth material that is not part of recognizable living organisms. Over 1,000 million tons of organic substances are created every day by photosynthesis in green plants.

Heavenly beings from other planetary origins of an advanced culture evidently established space colonies that eventually flourished into living planets, just as we plant plants in a garden; and when the planets disintegrated into excessive animal and people population that vitiated it in proportion to the plant life, the earth and other planets were forced into a rest period, as now happens on our earth, just as a farmer must let some of his fields rest to establish restored fertility for plant life by field rotation periods. In fact, the universe around us grows so fast, that of all heavenly bodies nearest earth, Earth is the only one with recognizable life today, the other planets being in a wintering rest period from active life organisms as we know them. This corner of the vast cosmos shows considerable ignorance in maintaining planetary hygiene and health seeing that, - the ignorant farmer raising too many animals and seed crops that remove soil fertility without returning it back to the soil, - the only method of escaping home-made ecological calamities, is again "changing fields" or migrating to another space "island" where life can again be cultivated as happened in recent few hundred million years on earth. Thus, today space scientists have again drawn up plans to build space stations where mankind could migrate to escape this polluted, over-crowded and disintegrating earth. Man has only learned what was needed to support life the hard way, - by destroying these essentials.......Now he must station earth colonies in space where earth refugees must plant trees to grow earth's oxygen and atmosphere, rather than continue polluting the existing air, and grow plant foods, rather than live on recycled plant food or hogs, to have food in his larder. But this is again the difficult way around, like animals that will not breed into better species or conditions except when forced by man thru selection, confinement and artificial hygiene. If humans worldwide would only stop their blind destructive way of life, ever demanding freedom to pollute, overpopulate, and destroy air, water, plants, animals and humans by violent abuse and misbehavior, we would not be living surrounded with misery and suffering. Rather than going off into space to start rebuilding Paradise, and living a life that destroys itself faster than it can live and grow, why not start exemplifying and demanding the BUILDING OF THE PARADISIAN PERFECTION ON EARTH AT ONCE.

My task herewithin is NOT to elaborate on the wonders of scientific escapes which require thousands of generations to remedy, but I am summing up earth's life history and prehistory after the mechanistic and extra-terrestrial elements were transmuted into a Theocratic Society which willingly lived in strict observance of the Co-Creative Principles of Divine Design, THE AUGUST ORDER OF LIVING IMMORTALS IN CONSCIENTIOUS PRINCIPLES that started and guided mankind thru-out the millions of years of existence since first settling in Hyperborea.

In "THE RACES OF MANKIND", Calvin Kephart estimated the age of the earth to be 4.5 billion years (4,500,000,000 years), as substantiated by Doctors F. R. Houtermans of Bern, G. J. Wasserburg of the Calif. Institute of Technology, and G. H. Weatherell of the Carnegie Institute in Washington on Sept. 10, 1957, at the opening of the symposium on the earth's age before The International Union of Geodesy and Geophysics at Toronto Canada. This is also what is called 1,000 Mahayugas, one mahayuga being 4,320,000 which is divided into four Yugas: (Satya, Trata, Dwapara and Kali Yugas)

The earliest fossil forms of life described in "EVOLUTION OF THE BIOSPHERE" by Kamshilov lists Catarchean (2700 to 3500 million years) Archean (1900 to 2700 million) and Proterozoic subdivided into the SAYAN (1500 to 1900 million), YENISEI (1200 to 1500) and SONII (570 to 1200 million). After these earliest single celled algae and bacteria appeared in the Yenisei Proterozoic era 1200 million years ago, multicellular forms are placed in the doctrine of the evolutionary theorist. Earth scientists have been developing food sources from green algae, especially "Chlorella" used in the Soviet Union, Poland, etc. to feed domestic animals which are said never to fall ill and develop excellently, and green algae is being added to foods in Japan and Czechoslavakia, beside having become a space traveler (Karuzina). In a 230 liter tank, Chlorella algae, within 24 hours gives 5.29 dry matter, suitable for human consumption, having 50% protein, 20% fat, 10% carbohydrates, 10% minerals, with Vitamins A, C, and the B complex. However most important is the fact that chlorella liberates 100 times as much oxygen as its own volume. This Tsiolkovsky holds could be used to create artificial atmosphere hereonin spaceflights. The Elohim, and angels whose lot fell on our earth, on another disintegrating heavenly body, probably did not feed from algae from streams or lakes of Sayan Mts. and Yenisei River directly, processing dried algae. Instead the Hyperborean settlers are known to have nourished from oxygen, water vapor and components of the atmosphere elaborated mostly by algae even today, beside airborn bacterial food. Spores of bacteria and molds will withstand rarification of the atmosphere, mold spores capable of growth being found at heights of 21 kilometers even. What matters is that they did survive, altho such tasteless fare and abstinence from bodily form and pleasure, gave life a lack of motive that took millions of years to disintegrate into the pleasure seeking and multiplication of the sons of God and their Paradise of every kind of fruit good to eat and pleasing to behold and eat thereof.

Having established the feasibility of migration thru basic work of the algae, I shall progress to the observation that altho algae seem to depend on water to exist, yet they make oxygen, the Life-Carrier in all earthly organisms. A great part of the earth's air, and consequently all Life is dependent on the algae of the ocean to give oxygen, breath or better named "biogen". Next, Hauschka would baptize Hydrogen with the name "Pyrogen", because it is a fire substance, the Leo or Sun substance, causing the sun to be a mass of atomic fire. It must have taken many billions of years to form the first land mass in the primal waters of the ocean, yet it was the algae and plants that performed part of the transformation into tangible elements. Now, when it is burned it gives off water, but to call it "Hydro-gen" is misleading since it is not "water substance" except for 11% of the water. The ancients pictured Life as a flame fed by water,- hydrogen is the flame and oxygen the carrier. A hundred miles up in the atmosphere, the air is 99.5% HYDROGEN, but next to earth it is only 0.2% hydrogen.

Stellar material is 82% Hydrogen, 18% Helium, and our sun is 87% Hydrogen and 13% Helium. (Speaking of Stellar, astral or star material both scientists and mystics would seem to be on common ground.) Uniting with oxygen, Hydrogen burns explosively, pyrogenic biogenesis thus resulting from solar life energy creation. While hydrogen is the lightest substance, oxygen is heavier than air, but in the form of ozone it rises to form the high ozone layer around earth which protects life on earth from harmful radiations, altho admitting the beneficial rays.

The Egg-Born Second Race contained their Solar, Astral or Stellar racial attributes from the First Race, in their Hydrogen Content, but it was the Living Water of plants that yield Oxygen that Vaporous Hyperborean androgyns first lived on, until tangible living water in the form of juicy fruits were developed. Just as important as plants were the moving living beings, men and animals, since they give off CARBON DIOXIDE. Plants depend on carbon dioxide and water to make carbohydrates. This is why the first humans and animals were gigantic so they could provide the necessary carbon-dioxide of corresponding ecological importance needed to sustain the earth with vegetation,- which in return cycle the plants turned into oxygenated atmosphere needed by humans and animals. The First Race being Astral Immigrants had bodies of ethereal substance, and thus were able to inhabit the heavenly bodies composed of 82 to 87% Hydrogen, being the progeny of Celestial Men, the Pitris, called Lunar Ancestors. There have been periods when water was poor in oxygen, which caused dead plant material to turn to peat and coal, rather than rot as it normally does. Now, during one year green plants consume 550 million tons of carbon dioxide, which is combined with 25,000 million tons of hydrogen, in the process of which 400,000 million tons of oxygen are liberated. Evidently the Carboniferous Period resulted from volcanic activity vitiating the atmosphere with earthy dust, making air-eating impractical. Today both rain water and well water contain earthy dust, altho well water has much more calcareous material.

In newborn babies the water content is 74% of the body weight, but in adults, wrinkled and dried by the absence of living water and stiffened by deposited mineral matter, they have only 50 to 60% water content. This shows that the younger, or more youth conserving, the higher its water content. Unless our way of life conserves this high water content, we do not stay young is another way to put it. Likewise, in the early ages of our Living Earth, in earth's Youth, it was mostly Living Water, and to this day, even after many ages of mineral deposits created by petrified dead plant corpses, the Earth's Surface is still 70% water. Water is the product of burning hydrogen with oxygen.

Actually, on the Paradisian Continent of Hyperborea, even the earth was covered with water after it was covered with fruit forests, green with vegetation which is nearly all living water, and while the bodies of the inhabitants were cloud-like vapor when they still nourished from vaporous air, as well as even when they were living on juicy fruits which gave solid bodies like the flesh of an apple but were more and purer water than babes bodies of our earthy species.

The Eternal Land of Hyperborea still in evidence is mostly in Siberia. It was named for a tribe, Sibirs or Sabirs, who once lived on the Ob River. Siberia covers 10 million sq. miles, equal to half the visible side of the moon, or the U.S. and Canada put together.

The Hyperboreans spoke of what now is called the RIVER OF THE FIRST PEOPLE or Forefathers, "Ensi-isa", now written Enisei or Yenisei. Along the river, near Krasnoyarsk Nature Reserve, one sees the unique Stolby pillar-like formation, appearing like ancient statues of our Forefathers, the First humans, staring out from the past in witness of a Life 100 million years ago. Yet, also in Krasnoyarsk, and not to be outdone by silent antiquity, is a modern and the world's largest hydro-electric plant which went into operation with a capacity of 6 million Kw.

THE RIVER OF OUR FOREFATHERS Like a crowd of ancient giants of ancient Hyperborea on a mount and reflected in the river in blue and violet shades.

Previously I have mentioned Lake Baikal, which means "Sunrise" Lake in Hyperborean, which science thus far has only been able to trace back 25 million years as having been formed by a huge earth crack. This we say comparing it with fossil remains from the Yenisei River and the Sayan Mts. which date a billion two hundred million and a billion nine hundred million years ago. Baikal contains ONE-FIFTH of all the fresh water reserve on earth, being the deepest lake in the world, or 5,710 feet deep, 400 miles long by 50 miles wide. In the summer it is navigable by large vessels, yet in winter it is crossed on ice. This is one of the sources of the Yenisei River, thru its Angara tributary. For every drop of melting snow water Baikal receives from surrounding high mountains, it yields a drop thru its outlet, much like the renown story of the lake of Galilee. The Baikal basin is fed by 333 rivers flowing into it. The immensity, calm and majesty of the Lake creates a deep feeling of reverence, so much that natives to this day will not let one word to be said against it, nor throw anything, not even a match into it. One can see clearly to a depth of 40 meters. The water is of a low mineral content and high ion content making it comparatively "live water" compared to distilled water, and is preferred where chemically pure "distilled" water is required such as in auto batteries. The Lake is an intense blue, claimed variously such as for its great age, high oxygen content of the water, and it is known to be older than the water in icebergs in Greenland. Yet, legend tells us that Lake Baikal became blue from the tears of a beautiful maiden who came there to weep for their betrothed in the mountains of Transbaikalia. Not only did the ancient Hyperboreans greet the sun, contemplating Illumination of this Living Earth in my pre-historic childhood millions of years ago, but today a rainbow of changing colors hails one there at dusk seen as the sun rolls down somewhere behind the spurs of the Altai and Sayan Mts.; it yields light last of all on the snows of Mt. Cherski, then the clouds turn crimson, the snow glimmers red, and fragments of ice sparkle like rubies. If the sky was clear, it too, from bright blue changed shades to crimson red, golden yellow and again back to azure and then finally darkness. But soon the star spangled sky beholds the Polar Star flashing like a diamond to the North. Nature's own monument to a Glorious Ancient Race.

In these most ancient territories, the flora and fauna are unique. Of the 1,800 kinds of plants and animals, 1,200 exist nowhere else. Hunting and fishing has been prohibited. Ducks stay permanently at their native lake, spending nights among ice hummocks. Salmon, pike, sturgeon, seal, etc. abound. Seven million of Siberia's ten million square miles is covered with tundra and taiga containing a magnificent variety of trees, from balsam popular to Baikal larch. Wherever my clan has dwelt, here in the mile high Andes, on Finn Hill across Lake Washington from Seattle, Finland and Estonia, as well as the Sayan and Altai Mts., the Yenisei R. and Lake Baikal, raspberries and other berries, along with apples, have accompanied us. This time I incarnated in a Finnish sauna that was next to a crab-apple swamp that was drained for a two-and-a-half acre vegetable garden, beside an orchard with most types of temperate fruit. The only tree unable to follow us in profusion was the birch, sparse in Western Washington, but relied upon in Finland, Altai, etc. Not only birch bark canoes, but like Essene bark sandals, the Finnish "Virsu" is a birch-bark shoe.

Often, I astrally return visiting the Altai, Sayan, etc. Sanctuaries to fellowship with my Immortal Brethren living hidden in caves of these precipitous mountains, much like the hibernating bear. Like the Arctic six months long day, alternated with the six months long night (altho the Altai Mts. are not in the Arctic Circle, but of same latitude as the northern border of the U.S.A.), one has to learn to exist on snow water and hibernate, as I was inspired on my six months fasts. But then comes the 6 months day, and the forest abounds with wild apples, currants, stone berries, raspberries, blackberries, birdcherries and many other fruits native to the Altai-Sayan Mt. foothills. Herbs of all kinds, bear's garlic, bulbs, etc. abound without needing to touch the squirrel and chipmunk's store of cedar-nuts and other seeds. Of course, one must have those fabulous "goat's feet" to jump across precipices and from rock to rock, like the pineal eyed Hyperboreans, as a song emphasizes: "To the high peaks, The far off places, Along paths that Donkeys cannot climb!" Explorers traveling in this region learn that deer are indispensible pack animals in the Taiga, but once in the mountains, other than very risky rafts that can crash to death in the rapids unless one knows the territory very well, there is no way to travel but like the wild goats, jumping from rock to rock, or crawling up and down cliff-like canyons.

The reader must realize, holding in mind, that people now inhabiting the land of the Ancient Hyperboreans are far from the ideals of those ancient Paradisians. Today, in the Eastern Sayans, at Tofalaria, we are told of a place "something like Tibet" ("Wilderness Survey" by Vladimir Chivilkhln). The whole nation adds up to 500. "They live like gods, high up in the sky...The militia and judge had no work in Tofalaria because no crimes had been committed there for centuries. The local doctor is bored to distraction from lack of patients. The Tofa families are sound and there are no divorces. The houses in Tofalaria have no locks. Tofa honesty is proverbial in the Sayans." The Sable Preserve guards took away the rifles of the surveyors, since shooting game is not allowed, regardless of bear, and other dangers.

But in the lower country, the older herdsmen will protest the eating of natural food: "Greens are for the reindeer, and the reindeer are for man"! Trees are fallen in the Taiga to obtain moss for the deer on pack trails since deer cannot go 3 days without moss, some of which is claimed edible for humans. But something of ancient reminiscence has been creeping in with modern technology. Thermal water springs abound around Lake Baikal, which are used for a cheap source of heat. Since 1969, a hothouse complex near Ulan Ude (S. end of Baikal) has been growing tomatoes, cucumbers, lettuce, onions, etc. practically all year round in spite of inclement winters. Even near the Arctic Circle, they have found that plants grow faster in such extreme North because the dark night never comes in the summer, so that more light falls on a square meter of ground in 24 hours than in South Siberia. This should be of interest to modern Paradisians, who do not wish to move south where hardships in pioneering are really greater due to insect or pest damage to plants being greater where there is no frost, the land available is poor, eroded and rugged, good land being taken and is contaminated by agro-chemicals, and with the population problems one's environs and habitants become hostile.

The predominantly juicy fruit diet, including succulent vegetable salads, is just as practical for northerners, as for those living in the tropics, if not more so for the northern mental and psychic development, since not only is the organism more adapted hereditably to temperate fruits, but breathing in familiar atmospheric prana, tending to familiar plants and farming procedures, etc. does not require a difficult hard-to-grasp adaptation to tropical environs which usually takes a generation.

Not everyone believes the theory that the earth's axis changed so as to account for the tropical flora and fauna being found in Arctic regions. Science in general tends to hold to the concept that goes it was warm at the poles because it was unbearably hot at the equator, so it was only after considerable cooling of the earth that man and animals migrated to tropical equatorial regions. Usually they link the dark skin and features of the races to hot climates and whiteness as being lack of solar heat. However, white races living in the tropics for generations do not usually get darker in practice. Also our holding to the axis changing theory is due to the suddenness of the catastrophe, in order to answer why animals, such as mammoths, were frozen with the grass they were eating still in their mouths, which is unlikely if the earth had gradually cooled off. Freezing cold in a warm tropical area could only come with a sudden cosmic change in events, such as a drastic change in the earth's axis. This could account for the ancient tradition as to the Sacred Lakes, frequented by the pious, especially the many hot springs, as mentioned of the Lake Baikal region, as well as volcanic warmth of the waters of Sayan and Altai Mts. In notes of the

"Kalevala" (Kirby) tells of the "Sampo", an unfailing food and money mill, legendary talisman of prosperity, was sought by inhabitants of Pohjola or Hyperborean regions, due to some change in the climate bringing on famine, and Kirby also mentions old Persian books telling of change from 9 months summer and 3 months winter, to 9 months winter and 3 months summer in northern Asia.

The Altai and Sayan Mts., the Yenisei to the Urals were rich in gold, silver and copper, of which the Finns were the first metal-working culture. No other Nordic nation of earliest epochs investigated in excavations of regions where they lived, had these metal utensils and instruments which identify Finnish culture. Herodotus told how they used paintings and effigies of griffins, etc. in an endeavor to scare away the raiders. The Mongolians soon became wise to the griffins and paid no attention to them, Kephart claims, taking their gold by force, causing subsequent migrations of the Finns to evade the Mongolian Huns, due to their Hyperborean Pacifist ideals. Thru-out this region, stone figures are found whose human features do not stand out in the broad daylight, but at sunrise and sunset, they turn human in features by design. In general, Finnic peoples were peaceful, and rather than co-exist with the evils of invasion and usurpers, they moved to where they would be respected, since wealth is for all who wish to work for it, and it is a crime to come to bloodshed over it. This Hyperborean instinct seemed inherited by my clan, giving my early association of gold with blood, recognizing preciousness with blood, sweat and tears brought by hard work, besides the dangers of coveting that which leads to bloodshed.

The reason why I have detailed my data on this region so extensively, in comparison to other earlier cultural regions, is because they all had earlier origins in Hyperborea, and especially in so-called occult or esoteric schools, teachers have added all kinds of misleading and falsely invented teachings in their lack of information on the subject. By this I mean, such claims that it was in the middle of the Gobi Desert, or in Tibet or even the Himalayas, that the eternal Hyperborean Shambhala, Heavenly Mountains of the Gods, Meru, etc. exist, including the claim that Agharta is actually a subterranean hollow in the inside of the earth sphere, with a mass of blazing radio-active material in its center. The latter view of Agharta can only exist in a world of spirits who lack a physical body, and since the visits to Agharta, made by the teachers thereof, were only with their astral bodies, Agharta's existence points to the Astral World or Plane of Phenomena, as well as of the First Tattva or Akashic Records. As explained earlier in this chapter, the earth is growing, increasing in circumference, so that life once did exist even nearer to its center, as well as wherever solid material is being formed, and thus Astral and Akashic records of its eternal past contain all of millions of past happenings related to all matter. Understanding this, we can describe most anything as occurring somewhere in matter at some time in Eternity.

Now, the reason that beings like ourselves at present cannot live inside the "hollow earth", the same as when man dives deep into the ocean, where the pressure would be too great, and also if we had nuclear mass burning in the middle of earth's presumed "hollow", it would kill all living beings with its radiation...The reason we are not destroyed by the sun's radioactive rays is due to the earth's ozone layer and distance (93 million miles) between the sun and earth. A radioactive mass gives off lethal radiations, and if large enough to be of use a few thousand mile away, the suspension of the ozone layer of protection becomes impossible for oxygen breathing beings. Only in dream worlds can we create without regard to physical facts of existence on earth today.

The Altai regions do have the Grace of Shamanism, a label that easily scares off investigators. The dwellers of Sham-Bha-La are located in the most inaccessible, precipitous and hidden Sanctuary locations camouflaged by natural surroundings, but we shall be cautious thus about certain description since modern technology is still seeking to unearth gold, silver, diamonds and other precious substances in the region even if it is fearfully deficient in cultivable land. Tibetan Lamaseries had little motive of invasion, but their destruction is history. "No master, however high and independent he might be, would have the right to divulge to the world the most time honored and archaic of the mysteries of the ancient college temples", H. P. B. has already warned us.

"According to Tibetan Tradition, the White Island (Shambhala, etc.) is in the only locality which escapes the general fate of other dwipas (lands, continents), and can be destroyed by neither fire nor water for it is the Eternal Land." (H. P. B.) The last of the mysteries of cyclic transformation shall come now that man has passed from ethereal to solid physical state, from spiritual to physical procreation, and is carried onward in the opposite arc of the cycle, when human progeny was created not begotten and woman knew no man. Northern Asia is as old as man, we repeat, old as the Second Hyperborean Race, and we have shown that the LARGEST INLAND FRESH WATER SEA which once surrounded the Sacred or White Island, all match the data we gave about Lake Baikal, the world's largest inland freshwater sea. Beside that confirmation of ancient manuscripts, also the Altai Mts. were known as the Heavenly Mts. never affected by submersion or fire in excessive solar heat, even if the equatorial sun did pass over it in the earliest ages. Likewise, Kephart observed that during the last ice age ages, ice covered North America extending from New York to the Pacific Ocean and as far south as St. Louis, and from the Pyrenees and Alps over Northern Europe. "It did not extend into Asia, except north of Urals in north-west corner of Siberia." The only explanation for this was that Central East Asia was too far from the ocean, which seems a dubious reason, other than the Eternal Land's own invulnerability.

Shambhala will betray all who look upon it,- rather than the exotic display of artistically carved temples of a sacred city,- one only sees a mass of jagged rugged quartz rock, like any of thousands of such peaks, mixed with grey and other shades of stone, in the mountain ranges around the world. In fact, one is not sure if one is contemplating the white snow and ice, or the milky to transparent quartz rock glistening in the sunshine, and from the plains beside the mountains, or even from close aerial flight above it, one cannot distinguish it from numerous jagged peaks, jutting upward in characteristic diversity. As we described before, the Hyperboreans were artists in creating forms that disappeared into the rocks before one's eyes in stone. Moreover, in the "Kalevala" it describes Finnish-Estonian-Siberian mythology telling of their gods creating animals and other things from pre-existing materials however incongruous. Likewise,

Finnic legends describe the power by which persons are transformed into trees, flowers, etc. for the purpose of concealment. This I must say because whatever is said about the forbidden and secret, soon attracts those, who, textbook in hand, go looking for the college temples of the masters, seeking to match descriptions. Much easier was it for searchers to find vast underground libraries buried under the Gobi desert, and temples extending from India to Scythia, as Mdme. Blavatsky and many others have in unselfish pioneering, but which were made increasingly objectified in personal gains and adapted to exemplify special schools and doctrines, turning the mystical into circus sideshows and disdain by their governments leading to utter destruction of all this, as in Tibet. Moreover, now would-be seekers have wandered from these esoteric origins to western branch schools which make the doctrines more lax and adaptable to the thousands and millions, rather than express real truth.

"We must not believe in a thing said merely because it was said; nor traditions because they have been handed down from antiquity; nor rumors, as such; nor writings by sages because sages wrote them; nor fancies that we may suspect to have inspired in us by a Deva (that is in presumed spiritual inspiration); nor inferences we may have made from some haphazard assumption we may have assumed; nor because of what seems an analogical necessity; nor on the mere authority of our teachers or masters. But we are to believe when the writing, doctrine or saying is corroborated by our own person and consciousness. For this I taught you not to believe merely because you heard, but when you believe of your consciousness, then act accordingly and abundantly." (Buddha)

The Lamas of Tibet, the Ancient Taoists, the Hindu Yogis, the Gnostic Essenes and other great schools, have been exoteric branches of this Central Asian source, which also established earlier schools in the Gobi and Tarim Basin cultures adjacent to the Altai, but when tried and tested, the searchers of all these doctrines were not found pure enough in motive and living habits to be able to reach the highest abode of the Supreme Guides of Earth in the Eternal Lands. This I found true of many of the teachers I have quoted in the past, as the accepted sources of Esoteric

Wisdom, yet my searching came from my early genetic background and life in the North Altai regions. Similar Sacred Rock Sanctuaries are found around the earth, carved out of rock, witness to thousands and millions of years: less than 4 decades ago I searched out hidden caverns built out of rock, beside tunnels found thru-out the Andes; They also exist in Mt. Carmel inhabited as college-temples and gardens of the Gnostic Essenes; Pr. Cherenzi-Lind, the Master Kut Humi lived in such a cavern at Shigatse Tibet, and described many just south in the foothills of the Altai; Wm. Goodell described the Master Jain Yogis of Mt. Abu, India who he said lived on a frugal diet of a handful of figs a day and slept "unnaturally in caves", among too many others to mention, many of which are now abandoned.

The Altai Labyrinths being the original establishing Sacred Sanctuaries of the Patriarchs guiding the destiny of earth, are truly esoteric temples of Light in that the transparent quartz and translucent milky quartz are illumined within, physically as well as spiritually, yet are inaccessible by the profane wanderer and searcher. Only those truly proven and capable are able to enter. Rather than oxygen consuming candles or lamps or bearing the darkness of opaque rock, the inclement weather by day is brightened by light and ventilated by air vents coming from the precipices enclosing their outside gardens. The floors are of warmth-yielding natural, soft and fragrant wood, and whenever mental sunshine is desired, rose colored quartz caverns were once available. Amethyst or violet blue or clear purple quartz also was used to overcome addictions to alcoholic beverages and worldly sensual pursuits, but initiation is no longer the work of inner sanctuaries, and must be completed in branch esoteric schools, since return to the world and other choices are not considered by the Highest Initiates.

The Essene fig, date and grape garden cultivations in hidden canyons have been described in their scrolls and history, as have the Taoist, Christian hermits, Himalayan Yogis, Tibetan Lamas, but all these were patterned from the original pattern or models from the Heavenly Mts., some of which obtained their original seedlings from them. Here the first apple, berries and other fruits and vegetables were cultivated on earth. As we have described, the near perpendicular precipices do yield weathered material into pockets in the canyons giving silt of rich rock dust, beside moss and lichens growing on rocks, which afford the richest soil for small gardens of few plants, vines and trees. Yet to get within the labyrinth sanctuaries with connecting passages, hermitages, cells, libraries and gardens, ingenious tests of perseverance and skill are required. Not everyone is able to dive thru turbulent rapids to find underwater entrances in icy cold water. Then again, too often the curious even among advanced initiates have ended in some canyon absent of vegetation, perhaps surrounded only with crumbling carnelian-like stone cast all around one like viands in an abattoir, seeing only their need to retrace their steps back to the world in spite of their mystical covetousness.

Their libraries tell the history of universes beside our tiny world thru millions of years and light years. The Greeks received much of their Wisdom from Atlantis, but also they are known to have investigated the origin of all Wisdom in Hyperborea. Zeus, son of Konos and Father of the gods and men was represented on the Acropolis of Argos by a gigantic statue which had two eyes in the face and one on top of the forehead, thus named Zeus Triopes, which describes Hyperborean features.

Before the change of climate from the warmth of the tropics to ice and snow in winters of the Altai region, needless to say, the hidden caverns had no purpose in the original Golden Age or Satya Yuga, since the Paradisians were in no need of hiding from a mean and evil world, as well as needing no such protection from the weather. One feature of these frugivorous giant Hyperboreans that the modern decadence of the biosphere has made impractical or unbearable all year round, was their complete comfort at all times living without clothing of any kind. Today, if one goes out into the forest or nature in Scandinavia, Canada, U.S.A., etc. in the summer one must be equipped with mosquito proof protection, repellent, etc., in spite of killing frosts in the winter. In "Maitreya" I also described how when we went into the tropical jungle we were greeted with 6 different types of insects with the most painful stinging bites attacking whatever part of the body was bare. It was such an antithesis, since a few years earlier I had sat in the in the warm Sauna, at Finn Hill home, or sunbathing on Lake Washington, as I dreamed of my early life millions of years ago in the Hyperborean Colony of Kara Usu Nor, living comfortably in fruit forests, naked and unashamed, which in turn intuitively inspired me to re-establish such ideals in a New Shambhala in the Land Closest to the Sun. The story has endured thru loss of much blood, sweat and tears, finding Nature very upset and hence unkind. Unless the air is contaminated with aerial spraying, as in Florida, in a warm climate, blood sucking insects follow one wherever there is vegetation. Like the air, pure and free of insect pests, the blossoming fruit trees were healthy, not covered with moss and other growths, as well as free from preying insects, due to the vitality of the virgin volcanic pulverized rock soil.

One can easily locate the Eternal Land, the Sacred Island of the First People by studying a Physical or Relief Map of the world, noting the outline of elevations over 5,000 feet in the Sayan, Tannu Ala, Plateau of Mongolia, and the highest peaks in the Altais reaching up to 15,266 feet above sea level. This is the referred to Sacred Island with the Heavenly Mts. which are described as north of the Himalayas, whose land mass over 10,000 ft. elevation extends up thru the Palmirs to the Tien Shan Mts. in an arc around the Tarim Basin, or the Continent of India, of whose northern shore was this Island. Notwithstanding the claims made for the Tarim Basin and other regions as being the original Eden, the Altai Territory also is of the richest black soil area of Siberia, with meadows and forest, and like Eden, four great Rivers begin here. But going back to the most primitive words in the Hyperborean tongue, the root word with an "Eden", with the suffix "ema" including "Maa" all refer to mother land, so Edenema means to thrive or flourish. This Eden, or bountiful mother-land is also called Paradiis in Hyperborean. "In the Sacred Mountain Meru, which is perpetually clothed in the rays of the sun, and whose lofty summit reaches into heaven, no sinful man can exist. A dreadful dragon guards it. It is adorned with many heavenly plants and trees, and IS WATERED BY FOUR RIVERS, from thence separating to flow in 4 directions": Hindu legend tells us as we have also quoted often from the Bible. This Meru and Paradise are one.

Chapter X: (photos 3, grandmother, mother, mother and brother john) GRANDMOTHER, MOTHER MARIA AND HER BROTHER JOHN YURMAN ARMUTARKUS,- PRENATAL GENETIC EXPERIENCE FROM HYPERBOREAN SHAMBHALA

The First People, the Hyperboreans, were believed to be born of one androgynous parent, having faculties of both sexes, by virgin birth. This became related to the belief that a Heavenly God, the Sun God, had become the fecundating essence and hence his was Isa (Father) or First Father (Esi-Isa) or racial trait, essence or seed sown by the Sun, or Paike, and the children's traits were determined by other heavenly bodies or constellations ruling at the time of birth. In time birth-giving was no longer androgynous, or an immaculate con-ception of any kind, and the religious temples no longer held to these high standards. Before marriage was instituted as a sacred rite, women were obliged to spend earlier unmated life serving at the temple till they were claimed by men who came to celebrate fertility rites therein, so that before being claimed, the children women had under the temple's auspices were mystically attributed to having been conceived by Heavenly Gods.

In the "Protevangeion" by First Bishop of Jerusalem, James, brother of Jesus Christ, in the complete uncensored version it states that the centenarian High Priest Zacharias, called the 7 undefiled virgins of the tribe of David who had served in the temple and they cast lots to see who got to spin the various colors of yarn,- Mary got the true purple. The Purple Robe or Garment is symbolic of an exalted station, and of the Ascended Sun God in the Mysteries, just as raiment (robe) means body. To confirm this the Protevangeion text affirms "THEN THE HIGH PREST KNEW MARY". This could only mean that he conceived an infant of exalted mission on earth with her, since he had just betrothed her to Joseph himself due to her marriageable age, or 14 years in those days. The esoteric meaning, is hidden today by paragraphing and punctuating to give a different meaning which it does not contain in Aramaic, and some translations even censored this origin of the Immaculate Conception realized in the Holy of the Holies, removing the part of this text that conflicts with the Roman Church's Virgin Birth Doctrine.

Thus, Zacharias, the High Priest of the Holy City of Light, became the father of John (so-called "Baptist"), the Lesser Hierophant, who became the Greater Hierophant, Iesous Chrestos, thru the Baptismal Rite of Purification, and who after the mock Crucifixion allegorical Initiation became the Exalted or Risen Sun God, Iesous Christos, Anointed Immortal God in Spirit. Physically, John or Ioannes is the Divine Theologian, Ioannes, which Herod calls him along with others, but his own Disciples and followers call him by his title of the Anointed Savior. (The Restored New Testament translated by James M. Pryse, explains this "Initiation of Ioannes" which renders the Apocalypse intelligible.)

As we recorded earlier, there came a time in the Lemurian epoch when the sexual faculties of self reproduction (reproducing one's own body perpetually as the self lived immortally) became grossly abused by seeking pleasure by adulterating the body's living water rarified essence with reproductive substance of plants, which led to the adultery of human seed of androgynes. In the climax man fell so low as to mix genes with beasts and other practices seeking to mutate human form into lesser beings of limited brain and consciousness, giving mankind beasts of burden and submissive races that were used as slaves.

The Tradition from Hyperborea, the Paradise of the Sun God and Heavenly Beings survived thru-out the ages, in resurrected forms but accommodated to the degraded animal passions we have described in each culture. The Sun God "Paike" (pronounced Baike) in the original Hyperborean language, is of the same root as "Paev" when the sun is visible or day. The sun, whether called a God or "not God", in Spirit and in matter, belief or science, still has to be credited with giving life, fertility, fruit and all things in life man enjoys. This became Bakchos or Bacchus of the Greeks, Egyptians, etc. "It is I who guide you; It is I who protects you and SAVES you; it is I who am the beginning and the end." Orpheus, son of Apollo, was the earliest bard, identical to the Finnish Vainamoinen, who charms the trees and rocks with his beautiful music played on a lyre, is said to have introduced the rites of Bakchos into Greece, where Bakchos became identified with Dionysus. Esoterically Pythagoreanism is rooted in Orphism. However, as most pure and pious things the meaning took on mock interpretations, so that books now refer to the initiates of the cult of Dionysus having to eat raw flesh to obtain union with God, and celebrations are described with wild music, dancing, wine, feasting and sexual orgy. What once was the beautiful, virtuous and living on Paradisian fruits, unclothed in the Paradiis of Paike, became a degrading, vile orgy in already ancient as well as modern civilizations. Wherever the faith of Bakchos went, it became a land flowing with the fruit of the vine, milk and honey. The wand, staff, rod or scepter of Bakchos could be turned into a serpent at will, or striking his rod on a rock it would become a fountain of water or wine, and many other Bakchos myths were later adopted for the Bible prophets. The original Paike (Bakchos) was an Illumining Sun God of Seership and Divine Inspiration, as well as Paradisian plenitude not defiled by men. Aurinko, another Hyperborean name for sun means the Dawning Sun, or the Goddess Aurora, Sister of Paike, which contains the English word for Inca, aura, aureole, auriferous, auric, and "Aurora Borealis", the Northern Lights, which are believed to be a collision of charged particles from the sun with the gases of the atmosphere.

Beside the Eugenic ideal of matching mature wisdom with youthful purity, such as Mary of 14 conceiving with Zacharias claimed to be a hundred years of age, there is another form of Mystical birth of Sun Gods celebrated by Egyptian, Inca and other ancient royalty, which was conception before betrothal between brothers and sisters, or between child and parent as was the case of Adam and Eve. This is now forbidden as unlawful incest today, but even with past religious strictness, no one knew how many eldest sons or daughters were conceived before the bride was allowed to go with the bridegroom. Ethically, birth conceived by brother and sister of noble traits (or child with parents), is the nearest approximation possible to Virgin Birth. In Virgin Birth, the immaculately conceived would conserve the same bodily traits as the Parents of the Virgin Mother. In the Virgin Birth theory of Jesus his physical traits would come from Ann and Joachim, mother and father of Mary. So to avoid this, Mary's mother, Ann, was also called to have conceived Mary without Joaquim's aid, to sustain the line of Holy Ghost fathers.

Now, in the conception occurring when brother and sister are the only possible parents, whether they had sexual intercourse or not, either way, the same genetic traits are given to the child, since the brother and sister, alone or together still have the same parental traits. The child has the same ancestors in either case. Even now, brother and sister complement one another in male and female traits, which originally were united in one person, which accounts for their double potentiality in size and faculties. With the division of androgynous beings into two half sexed beings of male and female traits, Nature was able to enable survival of our inferior race which abused and destroyed its complete Self-Perpetuating traits. This blemish manifests thru-out Nature today, which in the Paradisian origins did not occur, and which could be overcome if mankind would restore the Pristine Order of Paradisian Perfection. There would be complete continuity of the Higher Self, conscious of having lived millions of years as one Self, ever holding to the same noble traits of the August Order of Living Immortals.

The above explanation may help the reader understand my own case. Strange intuitive faculties seemed to have been awakened in my ancestry causing my grandparents to retrace our genetic origins, as if to speak out the need of Retracing the Hyperborean Origins of Race as I have done in this work. My own genetic birth among the Taevastan Finnic Race (or Heavenly Hyperboreans), and parents recent retracing of the journey of the original peoples by my parent's body giving pre-natal reminiscence of the last 100 million years of my life therein, is also visible in photographic qualities. My grandmother, mother, and John, my mother's brother who I was named for, thru God's Grace bestowed their likeness upon my traits of genetic continuity in life. The very clear blue eyes, blond hair and rosy white complexion is visible even in black and white photographs. Grandfather and Grandmother of both sides being identical, as illustrated above as to the Virgin Birth simile of Sun Gods, they were from the region of the Sacred Lakes Peipsi and Virts, or the city of Tartu, Estonia's University and cultural center. The origin of my genetic parents family name, Yurman, relates to first a syllable "Ur" meaning Light, as used in the ancient Sumerian city, and Urusalem, former name of Jerusalem meaning city of Light and Peace; Man as in Manda (Gnosis) Manas (Mind), or phonetically adding "Y", the composite meaning is Light-Mind or the Illumined, "Buddha". A-UR-INGO=Sungoddess.

However,- like in my former incarnation as John "Baptist",- to my Finn Hill neighbors, as a child I was Viruloojanni. The Viru prefix referred to "virtutama" which means to sprinkle, rinse or baptize as well as "virguma" or awaken, vivify or resurrect. So like Herod knowing John by his Baptist human traits, rather than seeing his life being of Spirit, he called Jesus Christ, "John who has risen again" and thus allegorically they honored him as the Risen Savior, while another man Simon was crucified in his stead to please the Jews by confusion. Likewise, a similar story of John Baptist (Viruloojani) happened in this life. (This may have an alternative interpretation, from Veri "blood" and Looja "Creator" as "Son of God" or Creator.)

The tracing of Taevastan Finnish origins of people in Finland and Estonia (if not Hungary, etc.) is described by the Anthropological work "Suomalaiset valimeren auringon alla" (Finnish life under the Mediterranean Sun) which shows they were the First Colonists on Krete and Thera which later became known for its Minoan culture, which was adopted by ancient Greece. We shall explain that after continuing our theme above.

To the Finns in general, Estonians are called "Viro-lainen", referring directly to "viru-tama" or John's Baptists, in the sense of the GNOSTIC FIRST CHRISTIANS OF ST. JOHN, of which the "KARAITES" possess the oldest manuscripts still teaching John's doctrine. Notice the word "KARA-ites", still holding to KARA USU NOR Hyperborean origins of Almighty Faith and Sacred Lake Purification Rites (see chapt. VII) reaching back over 100,000 million years. The Karaites spoke of the "Teacher of Righteousness" in speaking of Elias (Elijah) who John Baptist reincarnated in the same Mt. Carmel Brotherhood of Essenes. Karaite authors also refer to Philo's writings in Alexandrian traditions, and all of these are found in MAGHARIAN Scriptures. Tens of thousands Karaites still exist all over Russia today. The Nazarenes or Mandaens, the Gnostics of St. John were hostile to Judaism, Christianity and Islam, altho Islam regards John as the greatest of Prophets. The Damascus Document and the Dead Sea Scrolls refer to the

Book of Jubilees. The Calendar of Jubilees was adopted by the Magharians and the Karaites, so that their New Year begins and other sacred times fall on Wednesday. Barthelemy, Vaux and Kahle, beside other authorities claimed that THE MAGHARIANS WERE ACTUALLY THE ESSENES, (see "Dead Sea Scrolls" by M. Burrows). Burrows also quote the Dead Sea Scrolls, "God Loves the Knowledge of Wisdom", and reflects on the possibility that Gnostics, Essenes, Magharians and Karaites were all of the same source. Dupont Sommers also held, "Gnosis is one essential concept of the sect," that saving knowledge was believed to come by Revelation, and Reality was purely Spiritual, uncontaminated by matter, the soul being imprisoned in flesh bonds and the world. Burrow's affirmation that the Samaritans, Essenes and Covenanteers of the Dead Sea Scrolls were connected, which thus substantiates our thesis of Hyperborean origins.

The MAGHARIANS, which means the Men of the Caves where the Scrolls are found, are the Hungarian "Magyars", part of the Finno-Ugric people and language that settled in Phoenicia and Krete in prehistoric times, and later had to flee north to escape war and captivity. Thus, in the Hungarian (Bulgarian, etc. overlapping) they retained the "Magyar" name, but in Estonia, "Eestlane" (First People) was retained, and in Finland, it was "Suomalainen". (This will explain to students of Prof. Edmond B. Szekely's Essene School teachings, why, he from Hungarian origins, had such affinity, and facility working with Essene and Aramaic Scrolls.)

In "Suomalaiset Valimeren Auringon Alla", E. Vuorio relates that "From the Isle of Crete, the Finnish mercantile marines sailed to the Black Sea, and up the river Dniepr until Waldai. There the sailors pulled their boats over to the river Dvina, and sailed to the Baltic Sea. From there they transported amber, an article in great demand in Mediterranean countries. Later the Hellens (Greeks) conquered Crete and after that this Finnish tribe fled to Samaria and eventually travelling thru Russia to arrive and settle around the Gulf of Finland." One has to be aware that religious, political and historical authorities hid much evidence by inventing complicated ways of spelling words and adding different suffixes from one language to another. As we explained "Magharians" are simply Magyars, so the "Phoenicians" are thus phonetically the Finnish (Finnish-ians). Estonian, Hungarian and Finnish archeologists have found under the Grecian and later Phoenician cultures on Crete, etc., an even earlier Finnic racial and language culture. Kreteis the home of the Minoan culture that ruled over the East Mediterranean. The Palace of Knossos (Gnosis) believed by some to be the greatest wonder of the ancient world, is claimed to be of a peace-loving people who lived unfortified, and yet was destroyed about 1400 B.C. by volcanic ash.

Sir Arthur Evans did much to glorify Crete, restoring the Palace at Knossos. According to the "Secret of Crete" by Prof. Hans Georg Wunderlich, Sir Arthur actually overdid it, in making the statues and frescos unwarrantedly modernistic in the restorations. Wunderlich does concede that from Crete the Greeks obtained a new way of thinking, from the former cult of the dead (as in Egypt) to the cult of the Living. To the Egyptians the welfare of the dead was more important than that of the living, due to the fear of the hereafter. The Petrified Immortality of lavished tombs are but sterile mortuary palaces in Egypt. Thus the starting point of Greek culture and language was developed around the cult of the heroes, living memory of heroes attracted exemplification, and intelligence is communicated by speech and the written language. Wunderlich thought the women of Crete frescos bared their breasts as a "sign of mourning", as is known of Egyptian women, as also mentioned by Homer, while the Hebrews rent their garments. Instead of seeking the intuitive knowledge of the ancients, these scientists now match wits about what findings indicate and argue about false assumptions.

Yet, Crete was the cultural Motherland of Europe, just as symbolically the princess Europa became the wife of king Asterios, Queen Europa giving name to the European continent. Another bit of data is that the word "Candy" is from a town in Crete where fruits were and to this day are preserved by soaking them in crystalized honey. The soil of Crete was poor but they carried on intensive cultivation of olives, grapes and grains and had animals for milk and wool. As with the early Greeks, much of the food was eaten without cooking, the sun serving as fire.

As to the Minotaur, this like the Griffins of the Finns, had its purpose in creating fear among invaders. The need for beauty ran thru the lives of the Minoans. The Egyptian art was stiff and rigid, as if done off a stencil, while Cretans put life with seeming movement into plants. Also, men wearing loin cloth and adorned with gold and silver mingled with bare-breasted women clad in magnificent garments, handsome and carrying themselves proudly, along with running water and flowering plants were depicted in life-like frescos. Fresh water was piped in, baths plentiful, flush toilets and sewers that a man could walk upright in and which still function perfectly, show the Minoans to have been hygienically-minded, cleanliness surpassing people in the Indus Culture before the Aryans, thought by some to be of the same Pre-Sumerian Finno-Ugric Peoples. As with the Finns, also the Minoans illustrated equal rights among men and women, honoring gods and goddesses alike. Cretans are said to have a simple Nature Religion, built around a Magna Mater, Great Mother Nature Goddess. The ancient Philistines, still showing a giant Hyperborean statue of Samson, were originally Kretans that migrated to the mainland. At a time 2,000 years before Phoenicians introduced letters into Greece, writing by both linear and semi-pictorial signs existed in Krete (Kephart).

The Pre-Sumerian Leleges of Anatola had developed widespread commerce on the Mediterranean long prior to 3,000 B.C. In fact Kephart credits the Leleges with having introduced agriculture into Southern Sweden by way of the British Isles as early as 3,000 B.C. The Leleges were most known as the metal working culture, and thus their prospectors, beside merchants, magicians and religious leaders, all found it convenient to travel and spread their culture thus, beside the Finno-Ugric origin of Pre-Sumerians, Leleges of Anatola (later called Phrygia, Asia which is now Turkey), and Minoan Civilization of Crete, they formed colonies known as the Etruscan civilization of early Italy before Rome, the Basques along the Pyrenees of France and Spain, the British Isles, Scandinavia and as we said before Finland and Estonia. The Leleges were a kin to Phoenicians who followed them as the dominant sea trading power.

Now, Kephart (Races of Mankind) sustains the fact that Colonists from Anatola and Phoenicia merged with the people of

Ireland and Great Britain. The Eastern Mediterranean customs and dialectic aspects gave this early similarity to a Hebraic and Irish identity. Thus, Conor MacDari, writes: "The name Phoenician is but a formulated one, purposely misspelled and secretly applied to obscure the past history of the Irish race, previous to the 12th century...by which the Roman Church gained a foothold in Ireland. The name Phoenician, if spelled correctly, would be Finician meaning Sun Worshipper or follower of the Sun. The Irish word Fin, meaning son, family or tribe, true, fair, white, pleasant, a name for the Sun and an Irishman; hence the Finicians were an order or caste of devotees of the Sun." Well, I had a good friend named Finn, an Irishman, with whom I would tongue-in-cheek dispute who was the Real "Finn" and of Finnic origins. However, MacDari goes on to claim that "Persian or Turanian race", the Hindus, Norse, Aryans, Buddhists, etc. were Finns or Irishmen. His derivation of Bible words and terminology shows origins of Irish words in Hyperborean Finns which I have found so helpful in the study of Aramaic. Rome had the history of Eire (Ireland) submerged in the bed of the ocean under the name "Atlantis", after Irish Pope-Kings ruled the world, MacDari holds.

But then this leads to another possibility: In "The History of Atlantis", Lewis Spence holds that about the time of the date Plato gives for the sinking of Atlantis, a race of men known as Azilian-Tardenoisian invaded Southern France and Spain. the Le Mas dAzil cavern or tunnel in Pyrenean Dept. of Ariege reveals they were vegetarian-fruitarians for "Piette discovered the stones or husks of oak-acorns, haws, sloes, hazel nuts, chestnuts, cherries, prunes and walnuts. He also found a handful of barley seeds..." Yet, before them was another race, "that arrived in Europe at the close of the Ice Age, or 25,000 years ago and seemed to have wiped out the Neanderthal." This was the "gifted race, the Cro-Magnon, whose art is styled "Aurignacian". Take note of Aurignacian, rooting from the Finnish word for Sun, "Auringon" in description. The Cro-Magnon average height was 6 feet 2 inches and his brain case was extraordinarily large in capacity. Spence also quotes Prof. Osborn, "These Aurignacians were 'the Oalaeolithic Greeks; artistic observations and representation and a true sense of proportion and of beauty were instinctive with them from the beginning." I only wish I could quote the descriptions Diodorus Siculus gives about the race of warlike women, more ancient than those of Pontus in Asia, who inhabited an island called Hesperia, where grew the golden apples or oranges beside all sorts of tree fruits of Atlantis. But the reader can get the original book to do that.

In an interesting paper presented in the Scientific American (1977) Thomas W. Jacobson analyses "17,000 Years of Greek Prehistory", in which finds at Franchthi Cave go back to 20,000 B.C., show that people lived from wild pulses and animals they found by hunting, and it was only after the 6th millennium B.C. that agriculture appeared, with domesticated animals and food crops, wheat, barley and fruits like olive, fig and grapes. An obstacle from glacial advances over Europe kept man back culturally.

I have given the above studies as to the origins of various tribes, races and people, to show that Linguistic Origins do not coincide with Ethnic Origins, or that one cannot tell racial traits or color of skin by the language people talk. Black men in U.S.A. do not speak African. As to adding color to Turanian and Aryan people, which are words most often abused, let me quote Webster's Collegiate Dictionary. "ARYAN: A member of that Caucasian race one branch of which early occupied the Iranian Plateau, while another branch entered India and amalgamated with the primitive inhabitants; a member of the people which spoke the parent language from which the Indo-European languages are derived, loosely, an Aryan speaking individual of the Caucasian race." In turn, Webster states, "Caucasian should be distinguished from the Aryan,"- the Caucasians named for people of the Caucasus supposedly who were white. the inhabitants of Caucasus include Turanian and Aryan, and some say Dravidian. Webster defines "TURANIAN: Ural-Altaic people and languages collectively." Then it gives: "URAL-ALTAIC: Designating or pertaining to a great family of agglutinative languages or the peoples whose mother tongues are comprised in it. Physically these peoples vary from the pure Mongolian type of Eastern Siberia to the Caucasian Finn and Magyar." Formerly it was assumed that because the Finno-Ugric people gave linguistic traits and culture to the Chinese, that Finns were of the yellow race, and the Magyars were called

Hun-garians as if they had the Hun or Mongolian traits. All these confusions came with the evolution theory of races from the East Indies, as tho apes had become dark Dravidians and gradually migrating to cooler climates, man lost his dark tan to become white. It never takes into account that from pure white Paradisian peoples of former tropics lands of the North, man travelled away from a glacial cataclysm and taking on darkness mentally and in man's ways of life, he degenerated to lower cultures, and finally, beast. Apparently, "all the alphabets of the western world had their origin in symbols developed by the descendants of the early Turanian conquerors of western Asia. A corresponding relation probably exists with alphabets of the eastern world", wrote Kephart, after examining evidence that the Etruscans, Leleges of Anatola was identical to Runic inscriptions also used in Scandinavia, Germany and Britain as late as the Roman Conquest.

Having traced our ethnic and linguistic origins to a common fountainhead, I shall return to the story of my own Genetic-body's re-awakening, and retracing these genes within the trail of my grandparents. From 1892 to 1905 the Trans-Siberian Railroad made it possible for people near the Baltic to locate on both sides of the Ural Mts. Searching a better orientation, my grandparents thus ventured to settle in Russia, west of the Urals at Ufa in what then was the state of Samara, where the Zyrian or Komi Finnic people live, which again discloses the Karaite, etc. people fleeing from Syria and Samaria, which they renamed regions they had colonized phonetically. This is where my mother was born. Ufa is a mining and aircraft manufacturing center, but then a farming town with an old cathedral. However, when the railway reached Omski, East of the Urals, the family went there to take up homesteading land and building a ranch. They were living in these environs of the ancient Hyperboreans, witnessing the Sun rise from the East, over Lake Baikal, the Yenisei River, the Sayan Mts. and reach its Zenith after passing the Heavenly or Altai Mts. seemingly, working hard from dawn to dusk, probably not fully realizing why their Higher Mind attracted them there. Yet, early in childhood, I was told of the first occasion after the birth of my mother's brother, John, that grandfather and grandmother were in OMSK (where Om, -AUM abides) trading, and also 3 High Lamas (quite possible Their Holinesses, the Hambo Lamas of Ulan Ude, Russia and Mongolia) were passing thru on pilgrimage. Grandfather seeking diversion, had engaged the Russian speaking Lama in conversation, but grandmother being a pious reader of her Holy Bible in Estonian, was seeking to ennoble the coming Christ Kingdom. Not seeking to develop an adversary, the wise old Lama said, motioning toward the snow-capped Altai Mt. glittering in the sun southeast, "The river to this town flows from the summit of yonder Holy Mt. of the Gods, where the Adi Buddha first appeared on earth, and among your progeny (pointing to mother holding the babe, John), a great Lama (teacher), a Buddha in virtue, as well as an enlightened Teacher of your religion will be born in a New World, and lead our Faiths into One." The 3 Lamas were said to have continued their journey with a caravan doing trade across the Gobi for tea, etc. going on to Mongolia and China, but days afterward grandmother had a vision in which she seemed to be taken to the sunrise side of those Heavenly Mts., at the foot of whose Cathedral-like peaks of spires, there was a limpid lake surrounded by fruit trees and a race of happy inhabitants of a Paradisian land. The following year, it was confirmed that the Lamas had gone to visit the Sacred Lake on their caravan route, because of such prophetic visions and other omens occurring there, since the surroundings seem to be impregnated with (akashic impressions) the ancient happenings of a primordial glory. These events I have coordinated with research and data obtained from Pr. O. M. Cherenzi Lind the Tashi Hutulktu Budic Center at Tihwa, Sinkiang, on the south side of the Altais, who always included this Northernmost shrine among the Esoteric Sanctuaries on the letterheads he used, described as the "August Order of Living Immortals in Conscientious Tattwas".

This confided from my grandmother, contrary to the absence of any religious discussion, other than ridicule of church-attending Christians, so impressed my young mind, that I used to hunt for Finnish type words found thru-out Siberia in my geography, such as the mentioned "Tannu Tuva" (House of Thanksgiving), etc. My mother often recalled experiences such as the leg-tiring tramping of rammed earth building walls and privations of pioneering their homesteaded farm. People only ate flesh twice a year, from Livonian Knights indoctrination celebrating Christmas and Easter, holding more to the vegetarian Gnostic Essene doctrine of St. John's followers which they carried from Estonian origins, and even the Republic's flag represented people clothed in white linen Essene garments working black soil under the blue sky. They grew grain, mainly rye for bread, depended on clabbered milk from their cows, as well as cabbage used in sauerkraut, vegetables and cold climate fruits (currants, berries, apples, etc.) beside honey in place of sugar.

Like their hereditary forefathers had done for millenniums, they sought to cross the Western Ocean, Essene, Greek, etc. fabled Land beyond the Western Sea. They abandoned their years of hard work, this time traveling west on the Trans-Siberian railway to St. Petersburg, the capital of the Tsar Empire, then thru Estonia to London, and from there to America, then hailed as the land of opportunity, much like the Story of El Dorado (City of Golden streets of the Apocalypse) told of Incas. However, where they landed was the Brazilian mouth of the Amazon territory, where they only were met with tropical illnesses, beside fabulously fertile soil, with great mortality of European colonists. Karma again detoured the destiny of my genetic (subtle) body from going directly to the top of the High Andes, as now my grandparents continued with their children to the United States of America, where they settled again engaged in homestead farming, in the state of Montana near Chester.

The wide expanse of prairie required years of hard work with little gain. On top of this, mother received a hidden injury one day after hitching part-wild horses to a wagon and going to haul barrels of water home to the ranch house, of which she lost control, falling off the wagon, so that the desperate horses circled to come around and run over her with the loaded wagon. This gave a spinal injury, a kidney ailment, varicose veins and finally a hidden cancer that medical doctors never discovered until she had passed on. Yet, outwardly she was the very picture of health, so doctors could never understand her complaints

While she complained that she was brought up like a cat in a bag, never able to see the world for herself, in turn the younger brother, John, who she had raised since a baby, lived a life of breath-taking adventure. When young, near the time when

World War I started, he went to work in Wyoming where he had to live frugally to save any money existing mainly on watermelon, supplemented with bread now and then. When he returned home he greeted the family with a once round face and body that now had turned long and thin reaching over six feet in stature. It was then when my genes were given a physical body in conception, and due to the childhood and lives seemingly as soul mates in relationship, mother and John's love blossomed into my genetic traits, evident to all who see the photographs of the 3 of us.

However, another John, whom the Yurman family met in their travel to America, as grandfather whiled away time on ship from London to Brazil and the U.S.A. at the passenger's bar, who by this time had obtained a small farm tract he was developing, and sought a wife. Grandfather had casually suggested a solution on ship, and after 30 years of hard work on farms, mother sought to evade destiny, only by again going into bondage thru parental solution. Thus it was that this genetic continuity of body was "born" in a Finnish Sauna on Finn Hill, across Lake Washington from Seattle after the end of World War I. Only from family quarrels was it revealed to me that I was really "Yurmanijaani" and not "Virloojaani" as the Finnish neighbor's called me, or appearing to be the son of mother's husband, when he would reveal I was not his.

As to my mother's other brothers, Mike became established on a wheat farm in Montana near Chester, raising 3 girls, while Carl also married and established there. In turn, grandfather and his eldest son Andrew were disappointed about life in America, and thus left grandmother and her other sons to return to their ranch life at Omsk homestead in Siberia. No sooner were they again well established farming than the Tsarist Empire was overthrown by the Leninist Revolution. They thus became members of the "Aristocracy" simply because they had labored from dawn to dusk pioneering until they owned their own ranch, but this made them the enemy of Socialism, since they did not want to give up their hard earned "wealth" to be shared with those landless because they drank all the money they earned, lived without working hard as nomads or thieves, etc. Their land was confiscated for distribution among the landless thieving shepherds, and they were herded into corrals like cattle and railroaded to forced labor in salt mines. Andrew's buxom wife who weighed over 300 lbs., but strong from heavy farm labor, was reduced to a hundredweight skeleton. They wrote us from the camp reminiscing of their former rich farm diet, especially plentiful with butter which now was worth its weight in gold, having to subsist on frugal rations of bread, water and forced labor, as they pleaded to us for a few dollars to provide their barest needs. Eventually we lost contact with them, unable to help them since all mail was opened to reveal help or escape plans, and somewhere in Siberia I may have relatives. We feared the worst about grandfather, altho others are probably doing well now that everyone has gone thru the leveling-off adjustment.

In turn, to continue the story of mother's brother John, he spent some years working as a Ranger in Yellowstone National Park, later descriptions of which I loved to listen to by the hour, which inspired an early attraction to living close to Nature within me. At dawn he had often watched the bears wrestling with one another playfully like young men working out at a gym. The bears would call at each of the camper's cabins begging for a handout, but they could also be seen catching fish in a river or teaching their young how to do it, and he had joined them eating blackberries. He had sent us photographs of his life at the Park yet I wonder about my description in "Lessons from the Life of Love-Wisdom" telling of having lived on watermelon migrating along the Nile, which compares factually in pre-natal traits possible received from John's life in Wyoming working on a watermelon diet we told of above.

From these Glacier Park scenes where John spent his honeymoon after marrying Eva who was of Swedish descent, they went to live in Minnesota where they established the General Store for the farming community of Westbury. The prairie farmers would all come to their store to receive mail, deposit money in their U.S. Postal Bank, buy groceries, hardware, clothing, buy bakery bread and even have their auto and farm machinery serviced or repaired. One can imagine how they must have been up much of the night trying to do all this with basically only two people. One frosty morning, lighting a gasoline heater, it exploded, John tried to throw the burning mass out of the door, but only tipped it to an irremediable result that burned down the "town" of Westbury leaving them homeless. Recovering from this, they decided to go elsewhere.

They arrived at Juanita Grade School one afternoon, where I could direct them as they took me one in a little coupe practically without other belongings. After helping repair our farm vehicles, they leased a service station, with cabins and a garage on the north end of Seattle, working in the day, and everywhere they went they were in demand due to their artistic skill playing old time music in the evenings. Eva played the piano accompanied by John on the violin, which I so yearned to learn to play, and yet after much effort failed, not having entertaining-thru-music in my destiny. They continued in similar activities in the Puget Sound area till I left home. Then they invested in a new venture in Alaska, which included owning a schooner for hire, all of which went well except for an accident. John married again, and finally was able to retire in Arizona till his recent passing.

As to the origin of the inhabitants of South America who I later adopted as my neighbors here in Ecuador, this has such a varied possibility that I dare not make up my mind on any theory as yet. Most all who accept the existence of an ancient Atlantis, concede that its inhabitants peopled the Americas. There are many traits that resemble the Japanese among the Indians here which has given theories that they either crossed the Bering Straits to come here, or came across the South Pacific, since they less resemble Europeans and Africans what I may be able to contribute is clues to their cultural origins.

"THE OERA LINDA BOOK" was written by two people, the first of which was Adela, the seven foot wife of Apol, chief of the region called Friesland (the Netherlands). This book was a sacred volume that was passed on from generation to another in the Linda family for nearly 4,000 years written in rune-like letters. Adela tells of the sinking of the Old Land (Atland, Alt-land), which in the second part of the book is dated as occurring 2193 B.C., when it is believed the last land connections between Iceland, England, the Scandinavian Peninsula and Europe finally submerged. This was not Plato's Atlantis, remarks W. L. VALETTE (Fate, July, 1957. Vallette does say, "Atland may have been a last fragment of Aerve of Irish legends, the Cimmeria-Hyperborea of the Hellenes or the Ultima Thule itself. Soundings of up to 3 miles deep between Scotland, Norway and Denmark, make scholars believe that the land sank many thousands of years ago, tho they allow that the Frisian Islands were part of a larger mass which sank less than 4,000 years ago." (The reader might note that speaking about Leleges, Phoenicians and Kretans we said they traveled to the British Isles and South Sweden as early as 3,000 B.C.,- before this Atland event, and explains why the Finns pulled their boats over land a short ways between rivers connecting the Black with the Baltic Sea, rather than across Friesland).

Nearly 190 years after this Atland land-mass sinking (or 2,003 B. C. Woden, the son of a Friesian Sea King (Vi-King, Viking) became leader of all the "Children of Frya" who went to fight off the invasion of the "Children of Finda" in the East Baltic Sea, meaning the early Taevestan Finnish settlers of Estonia and South Finland. Woden had appointed Teunis as Neef or Admiral, with his cousin Inga under his command for thwarting the Finnish invasion. The Finns and their Predecessors were not warriors pirating ships or invading established properties, so they came to terms easily. Then the leaders separated: Teunis continued to the East, and Inga raised his blue flag on the sand inviting those among the Finns who wanted to join him to search for what could be found beyond the Frisian Sea (North Sea). Now, "Estoleland", Estonia, Essedon etc. are all of the same origin referring to the Eesti, as well as Atland or the Old Land of Hyperborean Continental origin, explaining how these people could communicate so readily, and why these Finns would want to join this Viking admiral to sail unknown seas in a search beyond for the lost remains of Hyperborean Atlantis, as they exchanged tales of their tribal origins. It was this tale of the Vast Arctic country called ESTOTILAND, once claimed to have no existence except in the brains of geographers, which consisted of New Foundland, Labrador and the area around the Hudson Bay, that caused the Cabots to sail from England in 1497 to rediscover and rename Estotiland, "New-Found-Land". Finn Magnesen established the fact that Columbus visited Iceland 15 years before his famous crossing of the Atlantic to (re)discover the New World. Eric the Red, Lief the Lucky and other Viking Norsemen actually started the first attempts of European colonization of Vin-land (East Canada) said to be named for grape-vine, but is a phonetic derivation of Fin-land, hundreds of years before Columbus. However, again the Irish claim they were the first, with Carbon 14 datings of iron-smelting furnaces in the Ohio Valley and elsewhere showing they were there in 600 B.C.

All this gives reason for the early Inca legend of blond, bearded white men who came from the Sunrise Sea and settled on Isla del Sol in Lake Titicaca, whose name was Inga or Lord-chief of the people of the Sunrise. The Araucanian (Mapuche) Indians retain tales of the Noble blood they received from Friesolandos (Frisians) who spoke the sacred language of the Inga and Finnish Noblemen. Even when the Spanish came to America, they were amazed to find that the blond, bearded teacher, Quetzalcoatl, as he was called in Mexico, had been there. In the Yucatan area he was called Kukulcan; in Guatemala, Gucumatz; in Colombia, Bohica, in Vene-Zue-la, Xue. In all those regions the early followers of John the Baptist (Gnostic-Essene-Chrestians) cult of Finnish colonists that joined Inga taught the Mysteries, obliging them to forsake their sins and their idols to worship the Heavenly Father, veneration of the Cross-Yoke, and baptism and fasting for the remission of sins. The word Inca in English, is pronounced Inga by the Indians of the high Andes, while the Quichua word "Mayana", not only refers to the Mayan people of Mexico, but is also the verb meaning "to wash" or baptize.

The "Isle of the Sun" in Lake Titicaca was settled by this Nordic colony of various dialects and yet similar Hyperborean origin. Thru-out centuries these white, blond and bearded became the established royal race, who had their own runic form of writing and language. Inca historians state that the Incas communicated with quippus, knots made in a cord depicting symbols, writing being almost unknown, except among the royalty who used quellas (books) of plantain leaves. The Royal Solar Race, as with other Solar Cultures, never married into the darker native peoples who became short, plump and with coarse black hair due to their almost exclusive corn, barley and fava bean diet. Only the sister of the Inca, as in other Sun worshippers rules,- was of high enough purity of blood by birth to be the Inca's wife. This principle of eugenic propagation is used by animal breeders to improve strains in their stock, yet humans are distressed by thinking it impure; certain European royalty were supposed to retain weak or unfortunate traits by marrying among those of close blood relationship. Bad traits come really by the similarity of abuses in living habits accustomed by privileged royalty, and it is foolish to blame evil blood in anyone when that comes by centuries of unwholesome living. Andean Indians may be Pre-Dravidians as in India.

Now, to elaborate about the Finns, Sedes, Danes, Latvians, Lithuanians, Prussians, etc. having close tribal origins in North East Asian Hyperborea, we should not be misled by languages which develop by the isolation of peoples from their origins. Tacitus described the inhabitants of the East Baltic Sea as the tribe of the Aestii (Eesti-Estonians), whose dress and customs are the same as those of the Suebi, but whose language resembles that of the British. "They even explore the sea and are the only people who gather amber, which by them is called glese and is collected among the shallows and upon the shore." The Suebi refers to the first Nordic Aryan nation in western Scythia which migrated to the Baltic Sea about 1600 B.C. and includes the Frisians, Swedes and Norwe-gians. The Norwegian invaders of Ireland were known as Finn-gaili, and since that means White-Heathen, Finns are simply whites, since they call themselves by their country "Suomi". In turn, "Suiones" are Swedes, referred to also by Tacitus, while the Finnish Lapps of northern Sweden are called "Finnae". The name Goths, from Gaut who was a great leader also gives our word for God, are more specifically the Lithuanians, Latvians, Prussians, Swedes, Danes, Norwegians. In their original home, and on the Baltic Sea they lived next to one another as one people until dialects in language and political boundaries now separate them from the Finnish peoples. The Language of the Livs (Latvians), Lithuanians and Prussians of North Poland has been modified by Iranian, Gothic, Estonian and Slavonian, and thus became known as the Vidivarii. There is a remarkable affinity between Sanskrit of the Hindu Kush and the Central Asian Getae who became the Lithuanians and other Gothic nations. The Balti, whom the Baltic Sea is named for, of Samarkand, became Rulers of Media, Zoroaster being of that family, and is believed to have given rise to Sanskrit language and culture. Germany is named for Hermanaric, who resembled Alexander as a warrior in conquest of a kingdom which extended from the Black Sea to beyond the Baltic. Herodotus is the first one to report that the Sarmatians who later inhabited Poland, "Polska" (home of the plains), descended from young Scythian men who married Amazon women of the Caucasus Mts., who spoke an impure dialect and allow their women to participate in warfare, so actually they differed little from other Goths of Getae. Nazarenes went to Hungary.

In, conclusion, we have sought to tell the mystical or spiritually oriented prehistory of mankind since our present human form became established on earth, and in this last chapter I have been prepared for the "Last Fight upholding the Banner of Shambhala". All know me as a champion of Peace and abstinence from bloodshed, esoterically knowing the "Sword of Maitreya" to be the Gnostic Serpent teaching: the WORD, or S--Word of the written instruction. I have traced genetic living records from the experiences of my grandparents in witness of Shambhala and the Heavenly Mts. in Altai area, what was ancient Hyperborea, and my parents custody of these traits, not alone racially as a Hyperborean, but actually genetically in traits harbored in my heart, from Hyperborea to the High Andes AQUARIAN AGE SHAMBHALA where the World Teacher was to appear. Likewise, already in my "MAITREYA, THE LOVEWISDOM AUTOBIOGRAPHY", in turn I traced the journey around the world, teaching Buddha Maitreya's arrival and establishment in the high Andes, by Pr. O. M. Cherenzi-Lind, Tashi Hutulktu, also of Altai's Northern Shambhala Shrine location. As the Tibetan prophecy reads, it is fulfilled: "When the Procession carrying the image of Shambhala shall pass thru lands of the Buddha and return to the first source, then shall arrive the time of the pronunciation of the Sacred Word Of Shambhala. Then shall the thought of Shambhala provide sustenance. Then shall affirmation of Shambhala become the beginning of all works and gratitude to Shambhala their end. The Great and small shall be filled with understanding of the Teaching. Solemnly I affirm, Shambhala the Invincible! Fulfilled is the circle of bearing the image! In the sites of Maitreya, is brought the Image, pronounced is Kalagai as the Banner of the Image unfolds...The Tashi Lama shall ask the Dalai Lama; What is predestined for the last Dalai

Lama?" He who denies shall be given over to justice and shall be forgotten, and the warriors shall march under the Banner of Maitreya. The City of Lhasa shall be obscured and deserted. To those in darkness, the Banner of Maitreya shall flow as blood, over the lands of the NEW WORLD. To those who have understood, as a red Sun... I shall create Satya Yuga and restore the Dharma to its former condition..."

Now, to substantiate the mystical or esoteric evidence, we would like to also strike a climax as to the findings concerning the Message of Maitreya on the Teaching of an ideal in a strict juicy fruit diet for the Paradisian Perfection of the New Race in the New Age of Aquarius which I have taught forty odd years. Health teachers and especially medical science have combated the ideas that man can live on only juicy fruit without using nuts, seeds, or some kind of "protein food". Yet, "Spiritualizing Dietetics," in spite of its errors allowing chemically toxic (sprayed, etc.) fruit, advocating distilled water purging in hopes of living without eating,- in the basic juicy fruit diet without seeds has reached approval as to the original diet of mankind by medical science. Fittingly, after two generations starting a third, on my so-called biological anniversary, July 23rd, 1979, a radio broadcast was given on the findings of Prof. Allan Walker of John Hopkins University of Baltimore, claiming man originally lived on a strict fruit diet. To substantiate these premises, Prof. Walker has found convincing proof in the 12 million year old remains of a human-like creature, or hominoid, which distinctly shows smooth unmarked enamel on the teeth, which could only exist if the diet consisted of juicy fruits. He pointed out that such hominoids definitely did not use nuts, roots or flesh of animals for food, which always show marked enamel where the teeth wear away due to tough fibers and substance that scratch the teeth. Grains, which not only are hard when eaten parched, etc. but also contain stone particles when ground with stone, are one of the worst markers of teeth, but they only date back to 10 to 15 thousand years. Cooking of foods also destroyed nutritional elements necessary for healthy teeth and body, as well created an unclean condition in the mouth that gave rise to caries which hardly exists in animals who eat no cooked foods.

The above new findings certainly upsets past theories as to the length of man's existence on earth, reaching unlimited possibility in existence, here or elsewhere as we would prefer to affirm, as well as establish the fact that humans belong on a juicy fruit diet. The Nutcracker man, or Zinjanthropus, of Prof. Leakey's findings in Africa, dated back to 1.75 million years, but this we have held to be evidence of the fall of man, acquiring the passionate, warring nature that led to development of the hairy apes and other species arrested in their cranial development and other animal traits. We hope the extreme age of 12 million years ago can be verified, which is 2 to 3 times older than the Ancient Gobi Culture, and contains more than two Maha Yugas. Satya Yuga, the Golden Age, associated with Hyperboreans, supposedly began about one Maha Yuga ago, or 4,320,000 mortal years ago.

The existence of fruit trees we have estimated at 100 million years ago, most probably having symbiotic dependence on frugivorous humans. The Soil and Environment has degenerated immensely since then, but the climax has come today when man has begun poisoning his food by poison, both plants and the soil. Thus, fruit eaters, already with chronic decay in their organism since childhood, and eating fruits grown in poisoned soil, find that their teeth rapidly develop caries in many cases. As to our advocacy of eating vegetable salads or greens to balance the excessive amount of sugar in sweet fruits, especially in the tropics, and obtaining nutrients necessary for rebuilding the body occasioned by past deficiencies, this we find more convenient for getting one's own source of pure food rapidly, than idealistic in what we seek to live on in the future, and the increased acceptability of good avocados, we find preference for the avocado dressing. And as mentioned earlier, avocados do have their disadvantage of being too heavy in fat, hard on the liver, which eventually should be replaced with the Fruit-Salad combination using bananas, papaya, and other fruits to make salad greens pleasant to eat. In summary, since one rarely has a taste for eating greens without dressing, having acquired a sufficient variety of fruits in one's garden, there is no reason to burden oneself with lowly vegetables and greens. Before, a neophyte comes up with any untoward conclusion, by this I do not mean one should eat fruit only if one does not grow it under one's own surveillance, but rather, the exclusive fruit diet and rewards in health with it, will come only when one's life is that of the PRISTINE ORDER OF PARADISIAN PERFECTION.

Shambhala: From Tropical Altai of the Gobi to Equatorial Andes!
In "THE LIFE OF BUDDHA" written by the 12th Patriarch, Asvaghosha Bodhisattva in the 1st Century A. D. we have the most admired account of the Buddha Sakyamuni, Gautama (the Victorious), the blue-eyed Prince, resplendent like the golden sun, who forsook all, along with his kingdom, for the Sake of the Dharma (Spiritual Law, Heavenly Kingdom).

Great sorrow arose in him contemplating the suffering seen everywhere, void of understanding, and yet ever seeking to add on more suffering by acts leading to suffering, old age, disease and death. After seeing so many pitiful sights in visits away from the palace, he withdrew from his royal retinue to a solitary spot at the foot of a ROSEAPPLE TREE (Syzygium Jambos) to sit and ponder all the untoward futility, whereupon, auspicious to the conducive and salubrious atmosphere of this Adi Tree, he entered in the Sublime Trance, Sam-Adi. "Pitiful indeed that these people who themselves are helpless and doomed to undergo illness, old age and destruction, should in ignorant blindness of their SELF-INTOXICATION, show so little respect for others who are likewise victims of old age, disease and death! I now will seek a Noble Law, unlike the worldly methods known to men. I will oppose disease, old age and death, and strive against mischief wrought by these on men." Seated there in increased Illumination, a Deva of the Pure Abode, who transformed himself into the shape of a Bhiksu, then appeared before him and said, "I am a Shaman, distressed and sad at the thought of age, disease and death... Therefore I search for happiness of something that decays not and never perishes, the joy of living in solitude of a lonely hermitage, untouched by any worldly source of pollution, free from molestation and all thoughts about the world destroyed." Then, before the Prince's eyes he disappeared into space, for he was a denizen of the heavenly regions, messenger of the hierarchy of former Buddhas. The Deva of the Pure Abode later enables the Prince to escape from his life at the Palace to abide in the forest for 7 years, eating little or no food, doing penance for evil deeds of the past, to a point where he realized he was at the opposite extreme.

Refreshed with the milk food offering of a maiden devotee, Buddha now seated himself under the Sacred FIG TREE (Ficus Religiosa), entering the transcendent State of Sambodhi. Under this tree, now called Bodhi Tree, of auspicious and conducive influence Gautama realized Buddhahood, Illumined among men."

Understanding the Gnosis of the text, and related references, we realize that the Pure Abode relates to the Paradise of SHAMBHALA, the home of Shambhus (venerable wise men) or Gods, and Maitreya Buddha.... Originally it was in the Heavenly or Altai Mts., called Meru also or Sumeru, the Abode of ADI BUDDHA, as well as the Java Aleim or Elohim as spoken of in the west. Auspicious traits accompany the attainment of the Buddhahood of Maitreya, who shall teach prophetically both the "Two Paths, the Doctrine of the Eye and the Doctrine of the Heart, intuitive symbols ever accompanying the Lovewisdom Message. The Dharma of the Eye embodies the Discipline in the external evanescent life, while the Dharma of the Heart embodies the Bodhi, the Permanent and Everlasting. wisdom turns the eye to the Heart, and Love sees all with Insight.

My travels in California foothills and mountains with desert-like climate foretold the eventual finding of isolated solitude from pollution of the worldly environs, not to be found in commercially valuable forest, fertile lands and wherever man can conglomerate already in 1938, and entering Ecuador, as told in my 1945 Autobiography I foresaw that this was only to be found in the warm, dry, Eternal Summertime climate of Tropical Andes Mt. Valleys of Southern Ecuador. However students lamenting that my Paradisian Discipline was hidden with my lone hermit Life in the hut of my Lake Quilotoa "Temple of Metta-Aum", embroiled me into returning to California, which made havoc with my discipline by poison-contaminated food source, beside confuse disciples as to seeking the Paradisian way of life and diet, nor were people ready to give up their own will and ways. Disappointed I returned to the Andes and finally was able to realize my permanent re-location in Loja Province of South Ecuador in 1962.

Everywhere I went in this region south of Loja, I found myself walking in lanes shaded from intense equatorial sun by benign rose apple trees (Yambos) whose fruit is called "Pomarrosa", and eating these rose flavored fruits was an experience "out of this world". It does not need much care, abounding in frost-free climates with enough sunshine, altho I had never seen it elsewhere in the many years I lived in Ecuador in its northern half. The flavor like fragrance of a rose is so up-lifting that one seems to partake of Paradise, or Shambhala, Meru or the abode of the Gods eating thereof. Here is a fruit that is not sour or acid to an extreme, nor is it insipidly sweet like others, crisp and buoyant when eaten. This seemed to be a fruit not only spiritually elevating to live on, but a secret key to longevity or immortality. Many of the inhabitants of the region were centenarians, which elsewhere I have attributed to low-protein and low-phosphorous soils and diet. The ROSEAPPLE has the same percentage of water as the common apple, or 84.5%, the low protein 0.6% along with 14.2% carbohydrates also resembles apples, and with 29 mg. of Calcium to 16 mg. of Phosphorus and 1.2 mg. iron (in 100 gr. portions) makes them unsurpassed as a Longevity fruit. It was most propitious that Gautama Buddha should be shown the way of Life he must lead under a rose apple tree bearing fruits of physical immortality as well as elevating the mind to ideals of True Immortality in Spirit.

Meru, abode of Brahma and Vishnu, is of the Seventh Zone of Spirituality, a Seventh Heaven, like the Sacred or White Island of unparalleled beauty, Cradle of Mankind which once was surrounded by sea, which now is the Gobi desert. As this region moved away from direct rays of the equatorial and tropical sun, northward, a Gobi Civilization prospered, followed by others, principally Jambu-dwipa, India, once part of the Lemurian continent which broke off traveling north to unite again with the Northern Hyperborean land mass. Thus, in ancient time this ancient continent of India was named for the Rose apple or Syzygium Jambos, called Jambu in Sanskrit. Pushkara, in the Puranas scripture, is also in the seventh zone, directly at the foot of Meru, and also is of two countries north and south of Meru shaped like a bow, which obviously describes North and South America. Meru, thus, corresponds to "Great High Mountain" which the Heavenly City of God, described in the Apocalypse of the N. T., which I have compared to be Mt. Chimborazo, actually the Highest Mt. on earth (2.5 miles farther from the earth-center than Mt. Everest of the Himalayas) in the Ecuadorian Andes named as the middle belt of the equator, and with 12 snow-capped peaks, glistening like pearly gates and whose inter-Andean Plateau resembles an avenue in translucent gold of direct rays of sunlight. In my "Maitreya" Autobiography I describe my calling to "the Land Closest to the Sun", Natural Mountain Temple, identical to Christian and Eastern Scriptural indications of Maitreya's domain.

Pushkara, or the Americas, are to be a prolongation of Jambu-dwipa whose center contains Meru, and altho this was true of India when it was connected with South America, Africa, Antarctica and Australia, it now blossoms in Truth in the Ecuadorian Andes. Jambu is near equivalent of Chamba phonetically, river I located on after coming to Vilcabamba, south Loja province, a Shrine of "Sacred Valley of Longevity" where thousands have visited because of its Paradisian message, tho not always with an understanding of its esoteric function. Shh...is thus more apropos to esoteric meaning in Shambha-La, rather than Jambu. Lord Buddha describing the "Jambu" (Rose-apple) land where the future Buddha describing the "Jambu" (Rose-apple) land where the future Buddha Maitreya will abide in Shambhala tells us: "Rich silken and other fabrics of various colors shoot forth from the trees. Trees will bear leaves, flowers and fruit simultaneously...A dragon tree will be the one under which Maitreya shall reach Enlightenment. The Rose apple (sketch at Left, caption: The Rose apple foretells a future) has another name "Eugenia Multiflora", just as the New Race (Eugenia) shall manifest the thousand blossomed chakra in Samadhi. Likewise the cotton-silk or Kapok tree resemble such a thousand petalled crown chakra and bears fabric used in blankets. Its thousand cone like thorns on its trunk appear like the hide of a dragon rather than a tree. Carob-family "faique" tree is another needle like thorn tree of our region, and the Sacred Fig numbers among the SIGNS OF MAITREYA and Shambhala in this unique land.

THE PREHISTORIC ORIGIN OF CONTINENTS ON OUR CHANGING EARTH

In this work we have told of the origin of continents from a single island, when the earth was "one vast watery desert", beginning with a skull cap, head-gear, which vegetation building, petrified; solid earth expanded to grow into a vast continent of the Eternal Land Hyperborea. After millions of years this broke from a southern part we called Lemuria. Geologists, studying their theories of drifting continental plates have come up with data co-inciding in fact to these records of ancient wise men, altho giving new names to what they seek exclusive credit.

Thus, they have named our Eternal Land, single continent, "Pangaea" (All Lands) which is surrounded by a vast ocean geologists call "Panthalassa" in an era 200 million years ago. 135 million years ago, they say this single continent split in two: The continental mass of North America-Europe-Asia being called Laurasia, and the land mass of India-Africa-Antarctica-Australia-South America being called Gondwana. The fossils of sheep-sized reptile called Lystrosaurus have been found in India, Africa, China and at the south pole Antarctica, showing they were all joined as adjacent territories in the age of reptiles 200 million years ago. Continents are shown to be moving upon a molten core.

When the molten rock of the earth's core cools and solidifies, it takes the magnetic poles direction of earth at the time. Thus, when land masses of continents move, the hard rock keeps the former direction of North and South Poles before continental movement, and this enables geologist to determine how continents are changing position. Of interest to earth's axis shifting theories is Paleomagnetic research that claims that the earth's magnetic field has reversed many times, flip-flopping from north to south 171 times in the past 76 million years. This sounds excessive, so I would inquire if this could not be the upheaval of clashing plates turning large chunks of earth mass upside-down, so the rock's poles become reversed, just as areas are found with geological earth strata formation completely reversed, oldest on top.

It was determined, however, that 450 million years ago the South Pole ice dome lay over what is now the Sahara desert. If we imagine this as the skull cap original island, and then expand its growth to reach the equator which then would have been the line of the 90 degrees East and West Longitudes of today, it gives a very rough idea of how the Antarctic continent, Siberia, etc. were on the earth's equator. this is our guess at present with continental shift data at hand. In time this original single mass spread further, till with the Antarctic as the south pole and Siberia near the north pole, the axis and continents have attained their present positions, with speculations as to future changes.

The hardened plates of the continental mass are like immense floating ice sheets, or rafts, 30 to 100 miles thick, moving on the molten hot earth mass, causing earthquakes where the solid crust rock breaks. At weak or thin spots holes break thru the earth's crust giving volcanic eruptions, especially where plates clash throwing up mountains. Thus, geologists claim that India broke off from Gondwana (which we call Lemuria) traveling north to crash into the Laurasian continent (Hyperborea) throwing up the Tibetan plateau and the Himalayan Mts.

Land at the extreme sides of the continents may be rising or sinking due to the continental plates creeping upon or under others, but the description of sunken continents may better be rendered as joining and separating traveling continents. Continents grew and divided, and these divisions mated with other parts forming still other ones on our Living Earth, imitating Nature and Cosmic processes. Lemuria did not sink entirely but rather the South American, African and Australian continents beside the Pacific Islands and India were divided from the single continent. Like Africa breaking loose from South America, North America broke off from Europe traveling west, so when it was in the middle of the Atlantic it was called Atlantis, until its mass got very distant from "just beyond the Pillars of Hercules" (Gibraltar) when it was renamed "America". Now, with the Poles of the Earth's axis in motion, the continents and oceans changing, and the races of man, animals and plants circulating upon the earth's surface, not only must the Earth Live, but also the Truth of Spiritual Oneness must be guiding this One to a higher and more Lofty Purpose. The Mundane Egg is the emblem of Eternity, infinitude, regeneration and rejuvenation, as well as of Wisdom. Endless in objective manifestation, shoreless in magnitude, boundless in space.

THE GROWING LIVING EARTH: Gave birth to the first dry land mass as an island which grew into a continent.

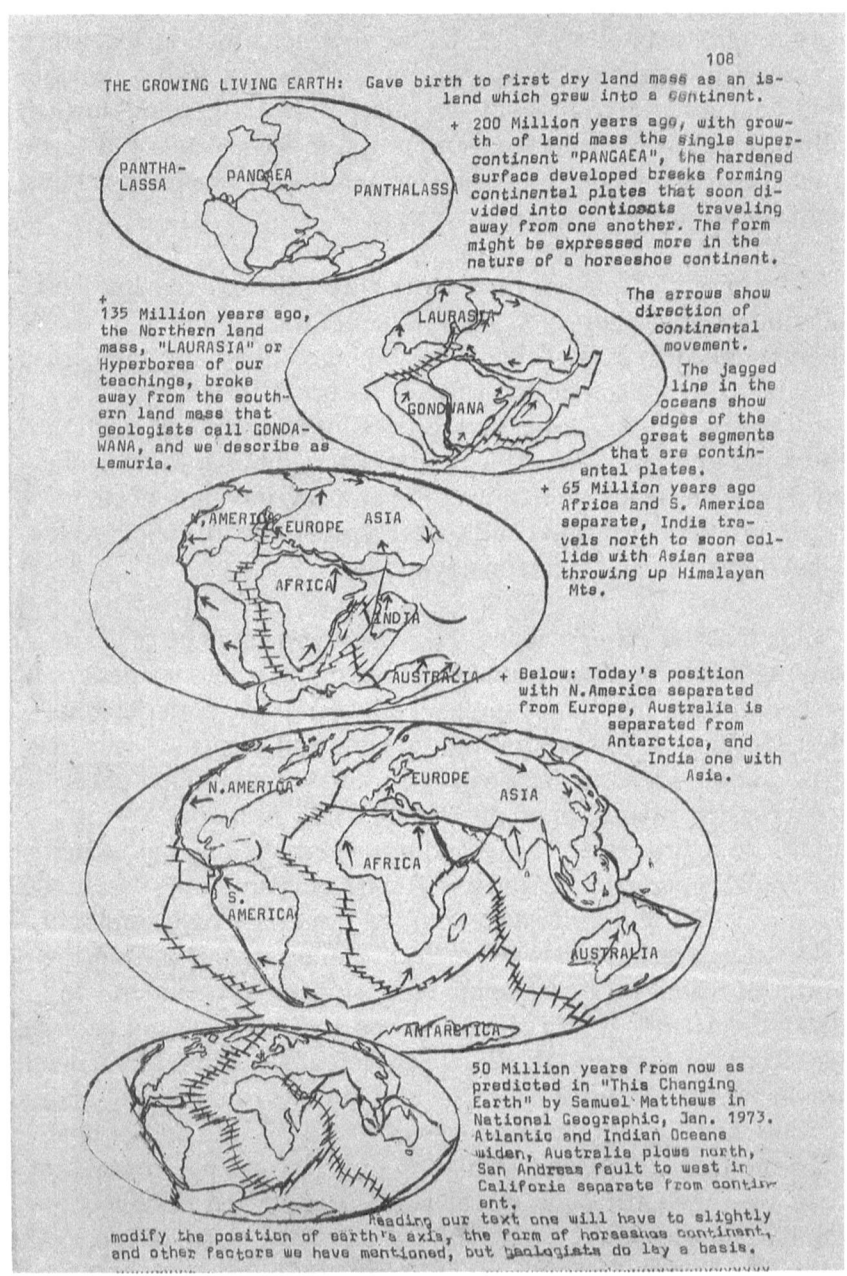

THE GROWING LIVING EARTH: Gave birth to first dry land mass as an island which grew into a continent.

+ 200 Million years ago, with growth of land mass the single supercontinent "PANGAEA", the hardened surface developed breaks forming continental plates that soon divided into continents traveling away from one another. The form might be expressed more in the nature of a horseshoe continent.

+ 135 Million years ago, the Northern land mass, "LAURASIA" or Hyperborea of our teachings, broke away from the southern land mass that geologists call GONDAWANA, and we describe as Lemuria.

The arrows show direction of continental movement.

The jagged line in the oceans show edges of the great segments that are continental plates.

+ 65 Million years ago Africa and S. America separate, India travels north to soon collide with Asian area throwing up Himalayan Mts.

+ Below: Today's position with N.America separated from Europe, Australia is separated from Antarctica, and India one with Asia.

50 Million years from now as predicted in "This Changing Earth" by Samuel Matthews in National Geographic, Jan. 1973. Atlantic and Indian Oceans widen, Australia plows north, San Andreas fault to west in Califoria separate from continent. Reading our text one will have to slightly modify the position of earth's axis, the form of horseshoe continent, and other factors we have mentioned, but geologists do lay a basis.

www.ingramcontent.com/pod-product-compliance
Lightning Source LLC
Chambersburg PA
CBHW030925180526
45163CB00002B/466